Selected Titles in This Series

724 **Lisa Carbone,** Non-uniform lattices on uniform trees, 2001
723 **Deborah M. King and John B. Strantzen,** Maximum entropy of cycles of even period, 2001
722 **Hernán Cendra, Jerrold E. Marsden, and Tudor S. Ratiu,** Lagrangian reduction by stages, 2001
721 **Ingrid C. Bauer,** Surfaces with $K^2 = 7$ and $p_g = 4$, 2001
720 **Palle E. T. Jorgensen,** Ruelle operators: Functions which are harmonic with respect to a transfer operator, 2001
719 **Steve Hofmann and John L. Lewis,** The Dirichlet problem for parabolic operators with singular drift terms, 2001
718 **Bernhard Lani-Wayda,** Wandering solutions of delay equations with sine-like feedback, 2001
717 **Ron Brown,** Frobenius groups and classical maximal orders, 2001
716 **John H. Palmieri,** Stable homotopy over the Steenrod algebra, 2001
715 **W. N. Everitt and L. Markus,** Multi-interval linear ordinary boundary value problems and complex symplectic algebra, 2001
714 **Earl Berkson, Jean Bourgain, and Aleksander Pełczynski,** Canonical Sobolev projections of weak type $(1, 1)$, 2001
713 **Dorina Mitrea, Marius Mitrea, and Michael Taylor,** Layer potentials, the Hodge Laplacian, and global boundary problems in nonsmooth Riemannian manifolds, 2001
712 **Raúl E. Curto and Woo Young Lee,** Joint hyponormality of Toeplitz pairs, 2001
711 **V. G. Kac, C. Martinez, and E. Zelmanov,** Graded simple Jordan superalgebras of growth one, 2001
710 **Brian Marcus and Selim Tuncel,** Resolving Markov chains onto Bernoulli shifts via positive polynomials, 2001
709 **B. V. Rajarama Bhat,** Cocylces of CCR flows, 2001
708 **William M. Kantor and Ákos Seress,** Black box classical groups, 2001
707 **Henning Krause,** The spectrum of a module category, 2001
706 **Jonathan Brundan, Richard Dipper, and Alexander Kleshchev,** Quantum Linear groups and representations of $GL_n(\mathbb{F}_q)$, 2001
705 **I. Moerdijk and J. J. C. Vermeulen,** Proper maps of toposes, 2000
704 **Jeff Hooper, Victor Snaith, and Min van Tran,** The second Chinburg conjecture for quaternion fields, 2000
703 **Erik Guentner, Nigel Higson, and Jody Trout,** Equivariant E-theory for C^*-algebras, 2000
702 **Ilijas Farah,** Analytic guotients: Theory of liftings for quotients over analytic ideals on the integers, 2000
701 **Paul Selick and Jie Wu,** On natural coalgebra decompositions of tensor algebras and loop suspensions, 2000
700 **Vicente Cortés,** A new construction of homogeneous quaternionic manifolds and related geometric structures, 2000
699 **Alexander Fel'shtyn,** Dynamical zeta functions, Nielsen theory and Reidemeister torsion, 2000
698 **Andrew R. Kustin,** Complexes associated to two vectors and a rectangular matrix, 2000
697 **Deguang Han and David R. Larson,** Frames, bases and group representations, 2000
696 **Donald J. Estep, Mats G. Larson, and Roy D. Williams,** Estimating the error of numerical solutions of systems of reaction-diffusion equations, 2000
695 **Vitaly Bergelson and Randall McCutcheon,** An ergodic IP polynomial Szemerédi theorem, 2000

(*Continued in the back of this publication*)

Non-Uniform Lattices on Uniform Trees

Memoirs
of the
American Mathematical Society

Number 724

Non-Uniform Lattices
on Uniform Trees

Lisa Carbone

July 2001 • Volume 152 • Number 724 (end of volume) • ISSN 0065-9266

American Mathematical Society
Providence, Rhode Island

2000 *Mathematics Subject Classification.* Primary 20–02; Secondary 22–02.

Library of Congress Cataloging-in-Publication Data

Carbone, Lisa, 1965–
  Non-uniform lattices on uniform trees / Lisa Carbone.
    p. cm. — (Memoirs of the American Mathematical Society, ISSN 0065-9266 ; no. 724)
  "July 2001, vol. 152, number 724 (end of volume)."
  Includes bibliographical references
  ISBN 0-8218-2721-9
  1. Lattice theory. 2. Trees (Graph theory) I. Series.
QA3 .A57 no. 724
[QA171.5]
510 s—dc21
[511.3′3]   2001023977

Memoirs of the American Mathematical Society

This journal is devoted entirely to research in pure and applied mathematics.

Subscription information. The 2001 subscription begins with volume 149 and consists of six mailings, each containing one or more numbers. Subscription prices for 2001 are $494 list, $395 institutional member. A late charge of 10% of the subscription price will be imposed on orders received from nonmembers after January 1 of the subscription year. Subscribers outside the United States and India must pay a postage surcharge of $31; subscribers in India must pay a postage surcharge of $43. Expedited delivery to destinations in North America $35; elsewhere $130. Each number may be ordered separately; *please specify number* when ordering an individual number. For prices and titles of recently released numbers, see the New Publications sections of the *Notices of the American Mathematical Society*.

Back number information. For back issues see the *AMS Catalog of Publications*.

Subscriptions and orders should be addressed to the American Mathematical Society, P. O. Box 845904, Boston, MA 02284-5904. *All orders must be accompanied by payment.* Other correspondence should be addressed to Box 6248, Providence, RI 02940-6248.

Copying and reprinting. Individual readers of this publication, and nonprofit libraries acting for them, are permitted to make fair use of the material, such as to copy a chapter for use in teaching or research. Permission is granted to quote brief passages from this publication in reviews, provided the customary acknowledgment of the source is given.

Republication, systematic copying, or multiple reproduction of any material in this publication is permitted only under license from the American Mathematical Society. Requests for such permission should be addressed to the Assistant to the Publisher, American Mathematical Society, P. O. Box 6248, Providence, Rhode Island 02940-6248. Requests can also be made by e-mail to reprint-permission@ams.org.

Memoirs of the American Mathematical Society is published bimonthly (each volume consisting usually of more than one number) by the American Mathematical Society at 201 Charles Street, Providence, RI 02904-2294. Periodicals postage paid at Providence, RI. Postmaster: Send address changes to Memoirs, American Mathematical Society, P. O. Box 6248, Providence, RI 02940-6248.

© 2001 by the American Mathematical Society. All rights reserved.
This publication is indexed in *Science Citation Index*®, *SciSearch*®, *Research Alert*®, *CompuMath Citation Index*®, *Current Contents*®/*Physical, Chemical & Earth Sciences*.
Printed in the United States of America.

∞ The paper used in this book is acid-free and falls within the guidelines established to ensure permanence and durability.
Visit the AMS home page at URL: http://www.ams.org/

10 9 8 7 6 5 4 3 2 1   06 05 04 03 02 01

CONTENTS

0. Introduction	1
1. Graphs of groups, tree actions and edge-indexed graphs	10
1.1 Graphs of groups	10
1.2 Group actions on trees and quotient graphs of groups	11
1.3 Edge-indexed graphs and their groupings	12
1.4 Existence of finite groupings	13
2. $Aut(X)$ and its discrete subgroups	16
2.1 Tree lattices	16
2.2 The group G_H of deck transformations	17
2.3 Constructing tree lattices	18
3. Existence of tree lattices	20
3.1 Locally compact groups and their lattices	20
3.2 Lattice Existence Theorem	21
3.3 Existence of non-uniform lattices on uniform trees	22
3.4 Existence of non-uniform coverings	23
4. Non-uniform coverings of indexed graphs with an arithmetic bridge	27
4.1 Geometric and arithmetic bridges in indexed graphs	27
4.2 Changing the ramification factor of an arithmetic bridge	30
4.3 Gluing unimodular subgraphs along connected intersections	31
4.4 Open fanning of arithmetic bridges	33
4.5 Indexed topological coverings	37
4.6 Step 1 - Schematic diagram	38
4.7 Step 2 - Construct topological covering	39
4.8 Step 3 - Change the ramification factor	41

4.9 Step 4 - Construct rectangles	43
4.10 Step 5 - Glue rectangles iteratively	44
4.11 Step 6 - Adjoin bridges	45
4.12 Step 7 - Multiple open fanning	55
4.13 Edge with a common factor implies non-uniform covering	68
5. Non-uniform coverings of indexed graphs with a separating edge	76
6. Non-uniform coverings of indexed graphs with a ramified loop	88
7. Eliminating multiple edges	94
7.1 Simplification of a graph with no loops	95
7.2 Graphs with multiplicities	96
7.3 Reduced factorization of an indexed graph	96
7.4 Canonical simplification of a unimodular indexed graph with no loops	97
8. Existence of arithmetic bridges	101
8.1 Unramified Loops	101
8.2 Completion	108
8.3 Suspension	109
8.4 Restriction	114
Bibliography	126

ABSTRACT

A *uniform tree* is a tree that covers a finite connected graph. Let X be any locally finite tree. Then $G = Aut(X)$ is a locally compact group. We show that if X is uniform, and if G is not discrete and acts minimally on X, then G contains *non-uniform lattices*; that is, discrete subgroups Γ for which $\Gamma \backslash G$ is not compact, yet carries a finite G-invariant measure. This proves a conjecture of H. Bass and A. Lubotzky ([BL]) for the existence of non-uniform lattices on uniform trees.

Our proof is constructive; we produce a non-uniform lattice Γ in G by constructing an infinite graph of finite groups with 'finite volume' which completely determines Γ. We first construct the 'edge-indexed' quotient graph $(A, i) = I(\Gamma \backslash\backslash X)$ of X modulo Γ satisfying certain necessary conditions, and we then obtain Γ as a finite 'grouping' of (A, i). This technique for constructing discrete groups which act on trees is naturally suggested by the Bass-Serre theory and was first proposed in [BK].

Our results show, in an explicit way, that the quotient graph $\Gamma \backslash X$ may have infinitely many cusps of arbitrary geometry, as is indicated by [BL]. The non-uniform tree lattices Γ are known *not* to have Kazhdan's property T ([VH]), to have arbitrarily large finite subgroups, are not virtually torsion free and cannot be finitely generated ([BL]).

The author was supported in part by NSF grant #DMS-9800604

For Hyman Bass on his 65th birthday.

Last night I dreamt of trees,
living, breathing, speaking,
showing me their secrets.
Their path in life to reach,
searching, finding, growing,
into infinity.

Mark Giglio

0. Introduction

Let G be a locally compact group, and μ a left invariant Haar measure on G. A discrete subgroup Γ of G is called a *G-lattice* if $\mu(\Gamma \backslash G)$ is finite, and a *uniform (or cocompact) G-lattice* if $\Gamma \backslash G$ is compact, *non-uniform* otherwise.

(1) In the early 1980's, Hyman Bass and Alex Lubotzky proposed to study lattices in the automorphism group of a locally finite tree X, a group that is naturally locally compact, in analogy with lattices in non-compact simple real Lie groups. While $G = Aut(X)$ is not simple, Tits has shown ([Ti]) that when G acts minimally on X, fixing no end of X, then G has a large simple normal subgroup, G^+, generated by all edge stabilizers. In fact, when X is homogeneous, G^+ is of index two, so G is 'almost simple'.

(2) The program of Bass and Lubotzky was motivated by the intermediate case of a simple algebraic K-group H, of K-rank 1, over a non-archimedan local field K, with finite residue field $\mathbb{F}_q$. The group $H \leq Aut(X)$ acts on its Bruhat-Tits tree X; for example, if $H = PSL_2(K)$ then X is the homogeneous tree X_{q+1}.

(3) With respect to this program of study, many of the natural questions have been treated. Some examples are: the existence of uniform tree lattices ([BK]), and of non-uniform tree lattices (the present work, and [BCR], [C2], [CR1]), the structure of uniform and non-uniform tree lattices ([BK], [BL]), covolumes ([BK], [BL], [IL], [R]), commensurability groups of uniform tree lattices ([BK], [YL]), super-rigidity ([LMZ], [BM]), the congruence subgroup problem for uniform lattices on regular trees ([Mo]), and the existence of towers of lattices ([CR2], [CR3], [R]).

(4) R. Kulkarni, in [K], has also indicated an analogy of the study of trees and their lattices with the study of discontinuous groups and Riemann surfaces, as well as direct connections with automorphisms of graphs, free groups and surfaces, and with the structure of finite groups.

(5) For the study of tree lattices in analogy with lattices in Lie groups, many interesting questions as yet remain open, such as the arithmeticity and commensurability of non-uniform tree lattices, the congruence subgroup problem on non-homogeneous trees, and

---

Received by the editor January 27, 1998

for non-uniform lattices, as well as direct connections between tree lattices and lattices in rank 1 Lie groups over non-archimedean local fields. We refer the reader to [L1] for a survey of the comparisons between tree lattices and lattices in Lie groups.

(6) The present work contains a proof of a conjecture about the existence of non-uniform lattices on 'uniform trees'. Bass and Lubotzky conjectured, in analogy with Borel's theorems in the classical case about the co-existence of uniform and non-uniform lattices in connected non-compact semisimple Lie groups ([Bo1], [Bo2]), that when uniform lattices are present in $G = Aut(X)$, under some natural assumptions, there should also be non-uniform lattices.

(7) In the case that X is the Bruhat-Tits tree of a rank 1 simple Lie group H over a non-archimedean local field K of characteristic $p > 0$, the existence of (both arithmetic and non-arithmetic) uniform and non-uniform lattices in H was established by Alex Lubotzky in [L2]. If K has characteristic zero, it is well known ([Ta]) that H cannot contain non-uniform lattices, while the existence of arithmetic uniform lattices in H was proven by Borel and Harder ([BH]). In [L2], Lubotzky showed that H contains non-arithmetic uniform lattices as well.

This work concerns the existence of non-uniform lattices in the automorphism group of a general locally finite tree. In order to discuss the precise statements and strategy, we introduce some terminology.

(8) Let X be a locally finite tree, and $G = Aut(X)$ its group of automorphisms. The stabilizers G_S of finite sets, $S \subset VX$, of vertices form a fundamental system of neighborhoods of the identity. In particular, the stabilizer G_x of a vertex $x \in VX$ is compact and open, in fact, profinite and hence totally disconnected. Thus $G = Aut(X)$ is locally compact and totally disconnected.

(9) A subgroup $\Gamma \leq G$ is *discrete* if and only if all vertex stabilizers Γ_x are finite. In this case, we define a volume:

$$Vol(\Gamma \backslash\backslash X) \quad := \sum_{x \in V(\Gamma \backslash X)} 1/|\Gamma_x|,$$

and call Γ an *X-lattice* if $Vol(\Gamma \backslash\backslash X)$ is finite, and a *uniform X-lattice* if the quotient graph $\Gamma \backslash X$ is finite. Thus a *non-uniform X-lattice* is a discrete subgroup Γ of G with infinite quotient graph $\Gamma\backslash X$ but *finite* covolume $Vol(\Gamma\backslash\backslash X)$.

(10) Locally finite trees which admit uniform lattices are exactly the universal covers of finite connected graphs; these are called *uniform trees* ([BK]). When X is uniform, a uniform (respectively non-uniform) X-lattice is a uniform (respectively non-uniform) G-lattice, and conversely ([BL]). We wish to understand when a uniform tree X admits a non-uniform X- (or G-) lattice.

(11) For X to admit a *uniform X-lattice*, it is necessary and sufficient that $G = Aut(X)$ be unimodular, and that $G\backslash X$ be finite ([BK]). Bass and Tits have shown ([BT]) that there are many uniform trees whose automorphism group G is discrete and hence is itself a uniform lattice, so it cannot contain a non-uniform lattice. Moreover, there are examples ([BL], and *cf.* section 3) where G is not discrete, but its action on X is not minimal (that is; there is a proper G-invariant subtree), and all X-lattices are uniform. We have the following, conjectured in an earlier version of [BL]:

(12) **Conjecture.** *Let X be a locally finite tree, and $G = Aut(X)$ its group of automorphisms. Suppose that G is unimodular and $G\backslash X$ is finite (thus, X admits a uniform X-lattice). If G is not discrete and acts minimally on X, then there is a non-uniform X-lattice Γ in G.*

(13) We present here a proof of conjecture 0.12. Our proof is 'constructive'; under the above conditions, we exhibit a non-uniform lattice Γ in G. Our strategy is to use a type of 'inverse Bass-Serre theory'; that is, we construct the quotient graph of groups $\Gamma\backslash\backslash X$ for the action of Γ on X.

(14) If $\mathbb{A} = (A, \mathcal{A})$ is a graph of groups with underlying graph A, vertex groups $\mathcal{A}_a$, edge groups $\mathcal{A}_e$ and monomorphisms $\alpha_e : \mathcal{A}_e \hookrightarrow \mathcal{A}_{\partial_0(e)}$, we put $i(e) = [\mathcal{A}_{\partial_0(e)} : \alpha_e \mathcal{A}_e]$ for each (oriented) edge $e \in EA$. The graph (A, i) consisting of A and the indices $i(e)$ labelling each $e \in EA$ is called the *edge-indexed* graph $(A, i) = I(\mathbb{A})$ for the graph of groups $\mathbb{A}$. An edge e is called *ramified* if $i(e) > 1$, and *unramified* otherwise.

(15) We form the graph of groups $G\backslash\backslash X$ for the action of $G = Aut(X)$ on X, and the corresponding edge-indexed quotient graph $(A, i) = I(G\backslash\backslash X)$.

(16) We restate our conditions on G and X in terms of the edge-indexed quotient graph

$(A,i) = I(G\backslash\backslash X)$.

(17) The existence of a *non-uniform X-lattice* $\Gamma \leq G$ then corresponds to the existence of an *infinite covering* $p : (B,j) \longrightarrow (A,i)$ of edge-indexed graphs (in the sense of 2.4) for some indexed graph (B,j), such that (B,j) has 'finite volume'. The non-uniform X-lattice Γ will be the fundamental group $\pi_1(\mathbb{B}, b_0)$ of a finite 'grouping' $\mathbb{B}$ of (B,j). Such finite groupings do not automatically exist, but [BK] give necessary and sufficient conditions for their existence. Namely, we require that (B,j) satisfy the combinatorial conditions 'unimodular' and 'bounded denominators' (see 1.4). Once these conditions are satisfied, there is automatically a finite 'cyclic' grouping of (B,j) (see 1.4), however, any finite ('faithful') grouping is allowable (see 1.3); this gives our technique for constructing lattices a great deal of flexibility. Finally, the fact that Γ is non-uniform is reflected in the condition that (B,j) is infinite.

(18) We will thus have proven the conjecture if we can exhibit infinite coverings $p : (B,j) \longrightarrow (A,i)$ with these properties for all the possible edge-indexed quotient graphs $(A,i) = I(G\backslash\backslash X)$ that can occur.

(19) The following theorem of Bass and Lubotzky ([BL]) indicates that the geometry of the quotient of a tree by a lattice is essentially arbitrary:

Theorem ([BL]). *Let A be any connected locally finite graph. Then there exists a locally finite tree X and an X-lattice Γ such that $A = \Gamma\backslash X$.* □

(20) We say that an indexed graph (A,i) is *minimal*, if (A,i) has no terminal vertices; that is, $deg_{(A,i)}(v) > 1$ for every $v \in VA$. In case (A,i) is minimal, we say that (A,i) is *non-discretely ramified*, if there exists $e \in EA$ such that $i(e) \geq 3$ or $i(e) = 2$ and $E_0(\partial_0(e))(= \{f \in EA \mid \partial_0(f) = \partial_0(e)\}) \neq \{e\}$; this is the criterion of Bass and Tits that ensures that the automorphism group of the covering tree is not discrete, (*cf.* section 3). The (combinatorial) statement of our main theorem is the following:

(21) Theorem. *Let (A,i) be any connected edge-indexed graph. Suppose that (A,i) is finite, unimodular, non-discretely ramified and minimal. Then (A,i) has a covering of edge-indexed graphs $p : (B,j) \longrightarrow (A,i)$ such that (B,j) is infinite, (hence p has infinite fibers), unimodular, has finite volume and bounded denominators.*

(22) Our proof is essentially as follows: we first prove Theorem 0.21 in the case that (A,i) contains an 'arithmetic bridge' β; that is, a set of $n \geq 2$ edges whose removal separates A into two connected components, and where the indices on the positively oriented end of the bridge all have a common factor $d > 1$ (section 4). In this case, there is a covering with the desired properties (see fig 0.22). A detailed description of the construction of such coverings, including examples, is given in [C3] (*cf.* section 4).

(23) We then prove Theorem 0.21 in the cases that (A,i) has a 'good' ramified separating edge (section 5), or has a ramified loop (section 6). In order to allow for the possibility that the indexed quotient contains multiple edges, we define a 'simplification' of a graph with no loops (section 7), which naturally has the property that for any edge e, the indices $i(e)$ and $i(\bar{e})$ are relatively prime. We also describe a method for treating unramified loops (section 8.1). In view of the cases treated, we can then assume that the quotient graph has no loops or multiple edges.

(24) To give a complete proof of conjecture 0.12, we need an 'existence theorem' for arithmetic bridges. The remainder of our argument roughly states that every indexed graph (A,i) with no loops or multiple edges, satisfying our original hypothesis must contain an arithmetic bridge or a 'good' separating edge.

(25) We prove this theorem in three stages: we embed (A,i) into the unique complete graph with the same vertices, and with new edges indexed so as to preserve unimodularity and relatively prime indices (section 8.2). We prove, using iterated 'suspensions', that the complete indexed graph must contain an arithmetic bridge (section 8.3), and finally, that the 'restriction' of the arithmetic bridge in the complete graph to the original graph yields an arithmetic bridge for the original (section 8.4). This sequence of arguments gives a complete proof of conjecture 0.12.

(26) We have obtained sufficient conditions for the existence of non-uniform lattices on a uniform tree X. In order to construct a non-uniform X-lattice, it is necessary to assume that $G = Aut(X)$ is unimodular and not discrete. The requirement that G acts minimally on X is sufficient but not necessary. Following ([BL]) we call X *rigid* if G is discrete, and we call X *minimal* if G acts minimally on X, that is, there is no proper

G-invariant subtree. If X is uniform then there is always a unique minimal G-invariant subtree $X_0 \subseteq X$ ([BL]). We call X *virtually rigid* if X_0 is rigid ([BL]). All lattices on virtually rigid trees must be uniform ([BL]). Conversely, in the case that X is not minimal, it is proven in [C2] that if X is uniform and not virtually rigid then G contains a non-uniform X-lattice Γ.

For $x \in VX$ we have $0 < \mu(G_x) < \infty$, where μ is a (left) Haar measure on G. When G is unimodular, $\mu(G_x)$ is constant on G-orbits, so we can define ([BL]):

$$\mu(G\backslash\backslash X) := \sum_{x \in V(G\backslash X)} \frac{1}{\mu(G_x)}.$$

In [BCR] we prove the '*Lattice Existence Theorem*', namely that G contains an X-lattice Γ if and only if G is unimodular and $\mu(G\backslash\backslash X) < \infty$. In particular, it is shown in [BCR] that if G is unimodular, $\mu(G\backslash\backslash X) < \infty$, and $G\backslash X$ is infinite, then G contains a (necessarily non-uniform) X-lattice Γ, which is a uniform G-lattice. For non-uniform G-lattices, in [CR1] we show that if X has more than one end, and if G contains a non-uniform X-lattice, then G contains a non-uniform G-lattice. In case X is a uniform tree, this result follows from the present work. Conversely, in [CC], we investigate the question as to the existence of non-uniform G-lattices in the case that X has a unique end, and G contains a non-uniform X-lattice.

(27) We have constructed a non-uniform X-lattice Γ by taking a *finite grouping* of our infinite edge-indexed graph (B, j), which is a 'non-uniform covering' of the indexed quotient graph $(A, i) = I(G\backslash\backslash X)$, where $G = Aut(X)$. Our construction is such that the indexed graph (B, j) satisfies [BK]'s necessary and sufficient conditions for the existence of finite groupings of (B, j), (that (B, j) be 'unimodular' and have 'bounded denominators' (see 1.4)). We have thus ensured that a finite *cyclic* grouping of (B, j) exists. The question as to *how many* finite faithful groupings an edge-indexed graph admits, up to isomorphism, is a complex and interesting question. In [CR2] and [CR3], it is shown that if X has more than one end, and if (B, j) admits a finite faithful grouping, then (B, j) admits an *infinite tower* of finite faithful groupings. This also implies that for the corresponding lattices, there is no lower bound on the covolume.

(28) With respect to the question of arithmeticity of tree lattices, we may adopt (see [BL], [L1]) a criterion of Margulis for detecting arithmeticity; that an X-lattice Γ is arithmetic if and only if its commensurability group $C_G(\Gamma) = \{g \in G \mid g\Gamma g^{-1} \smile_{comm} \Gamma\}$ is dense in $G = Aut(X)$. Y.S. Liu's *Uniform density theorem* ([YL]) states that if Γ is a uniform X-lattice, then $C_G(\Gamma)$ is dense in G, and so by this criterion, all uniform X-lattices are arithmetic. This is not surprising, as all uniform X-lattices are commensurable up to conjugation by an element of $G = Aut(X)$ ([BK]). It is known ([Mo2]) that the commensurability group $C_G(\Gamma)$ of $\Gamma = PSL_2(\mathbb{F}_q[t])$, the 'characteristic p modular group', is dense in $G = Aut(X_{q+1})$, where X_{q+1} is the Bruhat-Tits tree of $PSL_2(\mathbb{F}_q((t^{-1})))$. The group $PSL_2(\mathbb{F}_q[t])$ is well known to be an arithmetic non-uniform lattice, in the classical sense, as a subgroup of its ambient Lie group $PSL_2(\mathbb{F}_q((t^{-1})))$. We also know of examples ([BL], [Mo2]) of non-uniform X_{q+1}-lattices Φ such that $\Phi < G - H$, where $G = Aut(X_{q+1})$, $H = PSL_2(\mathbb{F}_q((t^{-1})))$ and $C_G(\Gamma)$ is discrete. The structure of the commensurability group of a general non-uniform X-lattice is not currently known.

(29) We survey some of the properties of the non-uniform lattices constructed here, and we refer the reader to [BL] for proofs and for further details. Let X be a locally finite tree, and let Γ be a non-uniform X-lattice. Then Γ is not finitely generated; the vertex stabilizers Γ_x, for $x \in X$, are arbitrarily large subgroups of Γ, so Γ cannot be virtually free, or equivalently, Γ cannot be virtually torsion free.

(30) It is known ([VH]) that automorphism groups of trees, and hence their lattices, do not satisfy Kazhdan's property T.

(31) The *Any quotient theorem* of ([BL], Thm 4.17) (*cf.* 0.19) indicates that any locally finite graph can occur as the quotient graph $\Gamma \backslash X$. Moreover, the number of 'cusps' of $\Gamma \backslash X$ may be finite or infinite, and $\Gamma \backslash X$ may have any geometric cusp structure that is 'combinatorially allowable' ([BL], 4.13). The rank of the free group $\pi_1(\Gamma \backslash X)$ may be any cardinal $\leq \aleph_0$ ([BL], 4.11).

(32) If X is the Bruhat-Tits tree of a rank 1 simple Lie group H over a non-archimedean local field K of characteristic $p > 0$, we know by Lubotzky's theorem ([L2]) that H contains uncountably many conjugacy classes of non-uniform lattices. By the non-

uniform existence theorem proven here, we know that there are non-uniform lattices inside $G = Aut(X)$ which contains H as a proper and 'relatively small' subgroup. Our construction of non-uniform X-lattices is combinatorial in nature, and will in general not place our lattices inside the Lie group H. By strengthening the coverings of edge-indexed quotient graphs developed here to *covering morphisms of quotient graphs of groups* with the desired properties, we can use a technique of Hyman Bass ([B]) to try to construct non-uniform lattices actually contained within H. We hope to address this elsewhere.

(33) In [BL], for homogeneous trees, and in [R], for uniform trees that admit non-uniform X-lattices, it is shown that for every real number $v > 0$, there is an X-lattice Γ such that $Vol(\Gamma \backslash\backslash X) = v$. Since the covolume of an X-lattice is constant on its conjugacy class, we may deduce that if X is a uniform tree that admits a non-uniform X-lattice, then X admits uncountably many conjugacy classes of non-uniform X-lattices.

The author is indebted to her PhD thesis adviser, Hyman Bass, whose careful attention and untiring efforts have played a substantial role in the development of this work. The author has great pleasure in thanking him.

Cambridge, Massachusetts, August 2000

NON-UNIFORM LATTICES ON UNIFORM TREES

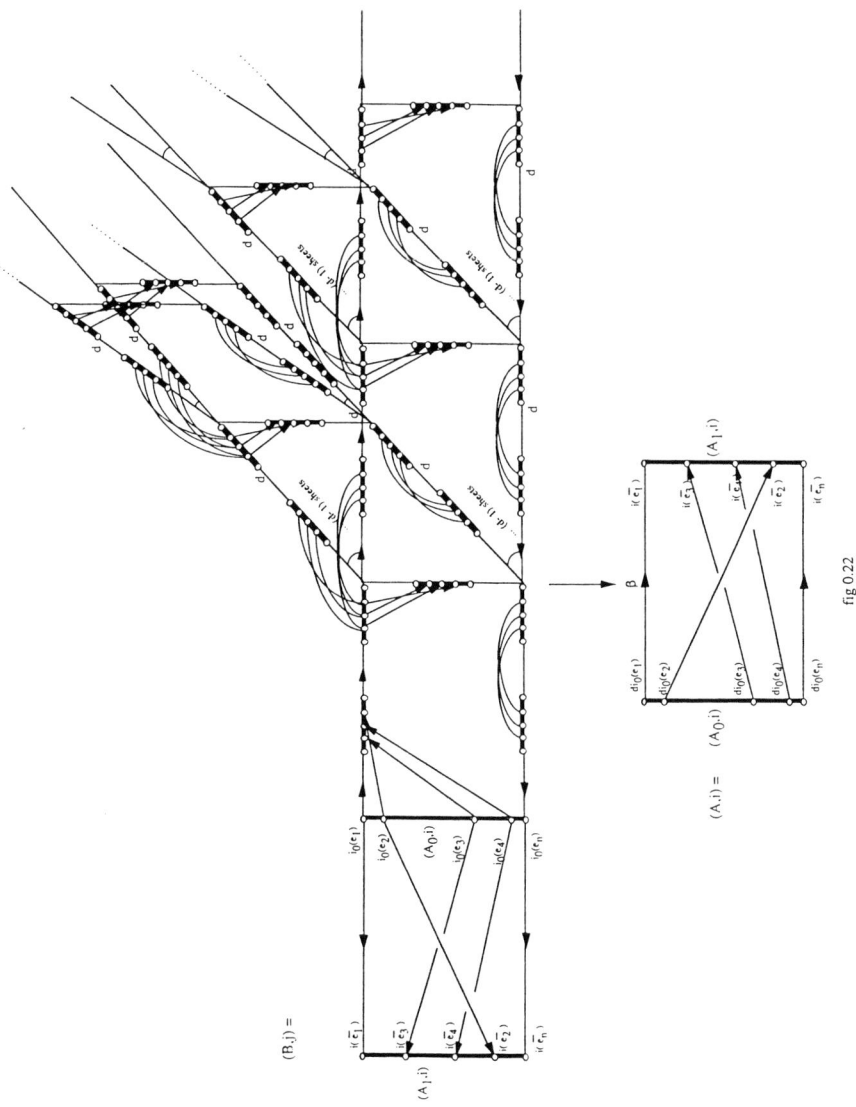

fig 0.22

The edge-indexed graph (B,j) for the non-uniform lattice Γ, and the covering $p:(B,j) \to (A,i)$.

1. Graphs of groups, tree actions and edge-indexed graphs

We shall assume that the reader is familiar with the basic aspects of the theory of group actions on simplicial trees as developed by H. Bass and J.P. Serre ([B], [S]). One of the essential ideas that emerges from this theory is that an action of a group on a tree is completely encoded in a 'quotient graph of groups' (*cf.* 1.2.3). This fundamental fact will allow us to construct tree lattices by constructing instead the appropriate graph of groups (*cf.* section 2.3).

An additional aspect of the Bass-Serre theory is that the 'edge-indexed graph' of a quotient graph of groups completely determines its universal covering tree up to isomorphism (*cf.* 1.2.3). As we shall see in section 2.3, this suggests a technique for constructing tree lattices, starting only with the appropriate edge-indexed graph.

To this end, we begin with the necessary preliminaries on graphs of groups and edge-indexed graphs. Additional references for this section are [BK], [BL].

1.1 Graphs of groups

Let $A = (VA, EA, \partial_0, \partial_1, -)$ denote a graph, with vertices VA, oriented edges EA, initial and terminal functions ∂_0 and ∂_1 that pick out the endpoints of an edge, and an involution, $-$, on the edge set that is fixed point free, and reverses the orientation. For a vertex $v \in VA$, let

$$E_0(v) = \{e \in EA \mid \partial_0 e = v\}.$$

Let $\mathbb{A} = (A, \mathcal{A})$ be a graph of groups, with vertex groups $(\mathcal{A}_a)_{a \in VA}$, edge groups $(\mathcal{A}_e = \mathcal{A}_{\bar{e}})_{e \in EA}$ and monomorphisms $\alpha_e: \mathcal{A}_e \to \mathcal{A}_{\partial_0 e}$.

Let $\Gamma = \pi_1(\mathbb{A}, a_0)$ be the fundamental group of $\mathbb{A}$ with respect to a base point $a_0 \in VA$ and let $X = \widetilde{(\mathbb{A}, a_0)}$ be its universal covering tree (see [B], [S]).

(1) A fundamental fact of the Bass-Serre theory is that Γ acts on X without inversion and with quotient projection $p: X \to A = \Gamma \backslash X$. Moreover, if $e \in EX$, $x = \partial_0 e$, $p(x) = a$ and $p(e) = f$ then $\alpha_f: \mathcal{A}_f \hookrightarrow \mathcal{A}_a$ is isomorphic to the inclusion $\Gamma_e \hookrightarrow \Gamma_x$.

(2) We call the graphs of groups $\mathbb{A}$ *faithful* if the action of Γ on X is faithful; that is, if

the defining homomorphism $\rho\colon \Gamma \to Aut(X)$ is injective. In general, any $\mathbb{A} = (A, \mathcal{A})$ has a 'faithful quotient' $\mathbb{A}' = (A, \mathcal{A}')$ with groups $(\mathcal{A}'_a)_{a \in VA}$ and $(\mathcal{A}'_e)_{e \in EA}$ which are quotients of $(\mathcal{A}_a)_{a \in VA}$ and $(\mathcal{A}_e = \mathcal{A}_{\bar{e}})_{e \in EA}$ respectively, so that the diagrams, for $\partial_0 e = a$,

$$\begin{array}{ccc} \mathcal{A}_e & \xrightarrow{\alpha_e} & \mathcal{A}_a \\ \downarrow & & \downarrow \\ \mathcal{A}'_e & \xrightarrow{\alpha'_e} & \mathcal{A}'_a \end{array}$$

commute and induce bijections:

$$\mathcal{A}_a/\alpha_e \mathcal{A}_e \to \mathcal{A}'_a/\alpha'_a \mathcal{A}'_e$$

(see [B], 1.24 - 1.25).

(3) It follows that $\mathbb{A}$ and $\mathbb{A}'$ have the same universal covering tree X. We conclude that if a graph of groups $\mathbb{A}$ is not faithful, we can always pass to its faithful quotient $\mathbb{A}'$.

(4) If $\mathbb{A}$ is a *graph of finite groups* $\mathcal{A}_a$, we can define the volume of $\mathbb{A}$ by

$$Vol(\mathbb{A}) = \sum_{a \in VA} \frac{1}{|\mathcal{A}_a|}.$$

1.2 Group actions on trees and quotient graphs of groups

(1) Let X be a tree and suppose that Γ acts on X without inversion; that is, for each $e \in EA$, $\Gamma \cdot e \neq \Gamma \cdot \bar{e}$.

In this case, we can form the quotient graph $p\colon X \to A = \Gamma \backslash X$, and we build a graph of groups

$$\mathbb{A} = (A, \mathcal{A}) = \Gamma \backslash\backslash X$$

on the quotient graph $A = \Gamma \backslash X$ so that for $e \in EX$, $x = \partial_0 e$, $p(x) = a$, and $p(e) = f$, the embedding $\alpha_f\colon \mathcal{A}_f \hookrightarrow \mathcal{A}_a$ is isomorphic to the inclusion $\Gamma_e \hookrightarrow \Gamma_x$.

(2) In particular,

$$[\mathcal{A}_a : \alpha_f \mathcal{A}_f] = [\Gamma_x : \Gamma_e] = |\Gamma_x \cdot e| = \left| p_{(x)}^{-1}(f) \right|.$$

(3) The fundamental theory of Bass and Serre tells us that

$$\Gamma \cong \pi_1(\mathbb{A}, a_0), \qquad a_0 \in VA,$$
$$X \cong \widetilde{(\mathbb{A}, a_0)}, \qquad a_0 \in VA,$$

that is, we can naturally identify Γ with the fundamental group of $\Gamma\backslash\backslash X$, and X with the universal covering tree of $\Gamma\backslash\backslash X$.

It follows that every graph of groups $\mathbb{A}$ encodes an action of a group $\pi_1(\mathbb{A}, a_0)$, $a_0 \in VA$ on a tree $X = \widetilde{(\mathbb{A}, a_0)}$, and conversely, every action of a group Γ on a tree X without inversion arises from a quotient graph of groups $\Gamma\backslash\backslash X$.

1.3 Edge-indexed graphs and their groupings

Let $\mathbb{A} = (A, \mathcal{A})$ be a graph of groups. For $e \in EA$, $\partial_0 e = a$, put

(1) $$i(e) = [\mathcal{A}_a : \alpha_e \mathcal{A}_e].$$

Thus i assigns a positive integer $i(e) > 0$ to each oriented edge $e \in EA$. We assume that all indices $i(e)$ are *finite*.

(2) If $i(e) > 1$, we say that e is a *ramified* edge. Otherwise, we say that e is *unramified*. We call $I(\mathbb{A}) = (A, i)$ the *edge-indexed graph* associated to $\mathbb{A}$, and we observe that every edge-indexed graph (A, i) arises in this way from a graph of groups ([B]).

(3) Given an edge-indexed graph (A, i), a graph of groups $\mathbb{A}$ such that $I(\mathbb{A}) = (A, i)$ is called a *grouping* of (A, i). We call $\mathbb{A}$ a *finite grouping* if the vertex groups $\mathcal{A}_a$ are finite and a *faithful grouping* if $\mathbb{A}$ is a faithful graph of groups.

(4) Let (A, i) be an edge-indexed graph and $\mathbb{A}$ a grouping of (A, i). Let $a_0 \in VA$ and put $X = \widetilde{(\mathbb{A}, a_0)}$. A fundamental observation is the following: ([B] 1.18, [BL] 2.5)

(5) $$X = \widetilde{(\mathbb{A}, a_0)} \text{ and the projection } p: X \to A$$

depend only on (A, i, a_0) and not on the grouping $\mathbb{A}$.

(6) We can therefore denote X as $X = \widetilde{(A, i, a_0)}$. We recall from 1.2.2 that p determines $[\mathcal{A}_{\partial_0 e} \colon \alpha_e \mathcal{A}_e] = i(e)$ for each $e \in EA$, thus p determines (A, i).

(7) In fact, if $x \in VX$, $p(x) = a$ and $e \in E_0(a)$ then we have the local map on vertices $p_{(x)} \colon E_0(x) \to E_0(a)$ and

$$i(e) = \mid p_{(x)}^{-1}(e) \mid.$$

(8) Let (A, i) be an edge-indexed graph and let $a_0 \in VA$. It follows that *every grouping* $\mathbb{A}$ of (A, i) gives rise to a group $\Gamma = \pi_1(\mathbb{A}, a_0)$ that acts on $X = \widetilde{(A, i, a_0)}$ without inversion and with quotient $p \colon X \to A = \Gamma \backslash X$. If we replace $\mathbb{A}$ with its faithful quotient we obtain a subgroup $\Gamma \leq G = Aut(X)$ whose stabilizers Γ_x and Γ_e are isomorphic to the vertex and edge groups $\mathcal{A}_a$ and $\mathcal{A}_e$ of $\mathbb{A}$.

For a vertex $a \in VA$, we define the *degree* of a in (A, i):

(9) $$deg_{(A,i)}(a) = \sum_{e \in E_0(a)} i(e).$$

If $\mathbb{A}$ is a finite grouping of (A, i), then we have ([BL] 2.6 (15)):

(10) $$Vol(\mathbb{A}) = \frac{1}{|\mathcal{A}_a|} Vol_a(A, i).$$

1.4 Existence of finite groupings

In this section, we describe the fundamental ingredient for constructing discrete groups which will be tree lattices. We shall start with an edge-indexed graph (A, i). A 'finite grouping' $\mathbb{A}$ of (A, i) will give rise to a discrete group $\pi_1(\mathbb{A}, a_0)$ which acts discretely on the tree $X = \widetilde{(A, i, a_0)}$. Our technique for constructing tree lattices will be described fully in 2.3.

Let (A, i) be an edge-indexed graph. The fundamental problem *'the grouping game'* is the following:

(1) Find $\mathbb{A}$ with $I(\mathbb{A}) = (A, i)$ and all groups *finite*.

where $I(\mathbb{A})$ denotes the assignment of 'group-theoretic' indices as in 1.3.1. In order to describe necessary and sufficient conditions for the existence of finite groupings on indexed graphs, we introduce the following notation:

(2) Let (A,i) be an edge indexed graph and let $e \in EA$. We set:

$$\boxed{\Delta(e) = \frac{i(\overline{e})}{i(e)}.}$$

(3) If $\gamma = (e_1, \ldots, e_n)$ is a path, set:

$$\boxed{\Delta(\gamma) = \Delta(e_1) \ldots \Delta(e_n).}$$

(4) We say that an indexed graph (A,i) is *unimodular* if $\Delta(\gamma) = 1$ for all closed paths γ in A.

(5) If (A,i) is a unimodular edge indexed graph and (A_0, i) is any proper subgraph of (A,i), then (A_0, i) is also unimodular, since closed paths γ in A_0 are also closed paths in A, and thus:

$$\Delta_{A_0}(\gamma) = \Delta_A(\gamma) = 1.$$

(6) Assume now that (A,i) is unimodular, and let $a_0 \in VA$. If γ is a path in A from a_0 to a, then $\Delta(\gamma)$ is independant of the choice of γ and we denote it $\dfrac{\Delta a}{\Delta a_0}$.

(7) We say that an indexed graph (A,i) has *bounded denominators* if the rational numbers

$$\{\frac{\Delta a}{\Delta a_0}, a \in VA\} \subset \mathbb{Q}$$

have bounded denominators.

(8) We observe that a finite indexed graph (A,i) automatically has bounded denominators.

The following theorem indicates that the conditions 'unimodular' and 'bounded denominators' on a indexed graph (A,i) are necessary and sufficient for the existence of a finite grouping $\mathbb{A}$ of (A,i):

(9) Theorem ([BK] 2.4) Finite grouping theorem. *An indexed graph (A, i) admits a finite grouping if and only if (A, i) is unimodular and has bounded denominators.*

In the notation of (7) above, we define the *volume* of an indexed graph (A, i) at a basepoint $a_0 \in VA$:

$$(10) \qquad Vol_{(a_0)}(A, i) = \sum_{a \in VA} \frac{1}{\left(\frac{\Delta a}{\Delta a_0}\right)} = \sum_{a \in VA} \left(\frac{\Delta a_0}{\Delta a}\right)$$

and we have the 'change of basepoint formula' ([BL] 2.6 (12)):

$$(11) \qquad Vol_{a_0'}(A, i) = \left(\frac{\Delta a_0'}{\Delta a_0}\right) Vol_{a_0}(A, i)$$

so the finiteness of the volume $Vol_{a_0}(A, i)$ is independent of the choice of basepoint.

2. Aut(X) and its discrete subgroups

Let X be a locally finite tree and $G = Aut(X)$ its group of automorphisms. The stabilizers

$$G_S = \{g \in G \mid gs = s, \ for \ all \ s \in S\}$$

of finite subsets $S \subset VX$ of vertices form a fundamental system of neighbourhoods of the identity. Moreover each G_S is profinite, hence compact, Hausdorff and totally disconnected. Thus G is locally compact, Hausdorff and totally disconnected. In particular, the vertex stabilizers

$$G_x = \{g \in G \mid gx = x, \ x \in VX\}$$

are compact and open in G.

2.1 Tree lattices

We shall see that under some natural assumptions, if X is a locally finite tree the locally compact group $G = Aut(X)$ contains *lattices*; that is, discrete subgroups of finite covolume. The action of G on X permits combinatorial criteria for subgroups $\Gamma \leq G$ to be discrete, uniform (cocompact), or to have finite covolume, without reference to a Haar measure on G.

A subgroup $\Gamma \leq G = Aut(X)$ is *discrete* if and only if

(1) $\qquad\qquad\qquad |\Gamma_x| < \infty$ for all $x \in X.$

In this case we can define a *volume*

(2) $$Vol(\Gamma\backslash\backslash X) = \sum_{x \in V(\Gamma\backslash X)} \frac{1}{|\Gamma_x|}.$$

We then call Γ an *X-lattice* if:

(3) $\qquad\qquad\qquad Vol(\Gamma\backslash\backslash X) < \infty$

and a *uniform X-lattice*, if

(4) $\qquad\qquad\qquad \Gamma\backslash X$ is finite.

(5) Thus a *non-uniform X-lattice* Γ is a discrete subgroup of G with infinite quotient graph $\Gamma\backslash X$ but *finite* covolume $Vol(\Gamma\backslash\backslash X)$.

(6) If X is the universal cover of a finite connected graph, it is shown in [BL] Ch1 (*cf.* section 3) that Γ is an X-lattice if and only if Γ is a lattice in the locally compact group G.

2.2 The group G_H of deck transformations

In this section, we describe some of the properties of the 'group of deck transformations' of a locally finite tree X; this is the subgroup of automorphisms that commute with projection. This group has the distinction that it will naturally contain the tree lattices that we are seeking to construct.

Let $H \leq G = Aut(X)$ be a subgroup without inversions and let $p\colon X \to H\backslash X$ be the quotient morphism. Let

(1) $$G_H = \{g \in G \mid p \circ g = p\}$$

be the subgroup of automorphisms that commute with projection to $H\backslash X$. We call G_H the *group of deck transformations* of X with respect to the subgroup $H \leq G = Aut(X)$.

The group G_H is the largest subgroup of G that has the same quotient as H:

(2) $$G_H\backslash X = H\backslash X.$$

We observe that G_H is a closed subgroup of G, hence

$$H \leq \overline{H} \leq G_H.$$

It follows that

(3) $$(A, i) = I(H\backslash\backslash X) = I(\overline{H}\backslash\backslash X) = I(G_H\backslash\backslash X).$$

Let $a \in VA$. We then identify X with $\widetilde{(A, i, a)}$, and we can write $G_H = G_{(A,i)}$.

Suppose that $\mathbb{A}$ is a faithful grouping of (A, i) and let $\Gamma = \pi_1(\mathbb{A}, a)$. Then $A = H\backslash X = \Gamma\backslash X$. Since Γ has the same quotient as H, the deck group G_H is equal to the deck group G_Γ. It follows that

$$\text{(4)} \qquad \Gamma \leq G_\Gamma = G_H.$$

Let (A, i) and (B, i) be indexed graphs and let $p\colon B \to A$ be a graph morphism. Then $p\colon (B, j) \to (A, i)$ is called a *covering* of edge-indexed graphs if for each $b \in VB$, $p(b) = a$ and $e \in E_0(a)$, the local map $p_{(b)}\colon E_0(b) \to E_0(a)$ satisfies

$$\text{(5)} \qquad i(e) = \sum_{f \in p_{(b)}^{-1}(e)} j(f).$$

In this case, we identify

$$\widetilde{(A, i, a)} = X = \widetilde{(B, j, b)}$$

so that the diagram of natural projections

$$\begin{array}{ccc} & X & \\ {\scriptstyle p_B}\swarrow & & \searrow{\scriptstyle p_A} \\ B & \xrightarrow{p} & A \end{array}$$

commutes.

(6) It follows ([BL] 3.3) that if $\mathbb{B}$ is a faithful grouping of (B, j), then $\pi_1(\mathbb{B}, b) \leq G_{(A,i)}$.

We shall see (section 3) that it is intrinsic to our technique for constructing non-uniform X-lattices Γ that Γ will be contained in a group of deck transformations G_H for a subgroup $H \leq G = Aut(X)$ without inversions.

2.3 Constructing tree lattices

The fundamental theory of Bass-Serre for groups acting on trees outlined in the previous sections gives us a method for constructing lattices on a locally finite tree X.

We recall that an action of a group Γ on a tree X is encoded by a quotient graph of groups $\mathbb{A} = \Gamma\backslash\backslash X$. We shall construct a lattice Γ by constructing instead the appropriate $\Gamma\backslash\backslash X$.

We begin with an indexed graph (A, i). Then by 1.2 and 1.3.4-1.3.7, (A, i) determines $X = \widetilde{(A, i, a_0)}$ up to isomorphism.

If we assume that (A, i) is unimodular and has bounded denominators (which is automatic if A if finite), by the finite grouping theorem, we can find a finite (faithful) grouping $\mathbb{A}$ such that $I(\mathbb{A}) = (A, i)$ and a group $\Gamma = \pi_1(\mathbb{A}, a_0) = \pi_1(\Gamma\backslash\backslash X, a_0)$ acting (faithfully) on X. Then

(1) Γ is discrete, since $\mathbb{A} = \Gamma\backslash\backslash X$ is a graph of *finite* groups.

(2) Γ is a lattice if and only if

$$Vol(\Gamma\backslash\backslash X) = Vol(\mathbb{A})(= \sum_{a \in VA} \frac{1}{|\mathcal{A}_a|} = \frac{1}{|\mathcal{A}_a|} Vol_a(A, i)) < \infty.$$

(3) Γ is a uniform lattice if and only if A is finite.

(4) Γ is a non-uniform lattice if and only if A is infinite.

This technique for constructing tree lattices was first proposed by [BK] and is the essence of the strategy we will undertake in the forthcoming chapters.

3. Existence of tree lattices

Let X be a locally finite tree and let $G = Aut(X)$. We wish to determine conditions that will ensure that G contains X-lattices. We first recall some facts about lattices in locally compact groups which will suggest some obvious necessary conditions.

3.1 Locally compact groups and their lattices

(1) Let H be a locally compact group, and μ a left-invariant Haar measure on H. Let $\Gamma \leq H$ be a discrete subgroup with quotient $p : H \longrightarrow \Gamma\backslash H$. The quotient morphism is locally measure preserving, so the quotient carries a measure $\mu(\Gamma\backslash H)$.

(2) We recall ([BL] Ch1), that Γ is called an H-*lattice* if $\mu(\Gamma\backslash H) < \infty$, and a *uniform* (or cocompact) lattice if $\Gamma\backslash H$ is compact, non-uniform otherwise. The group H is called *unimodular* if the Haar measure μ is both left and right-invariant.

(3) Let X be a left H-set with compact open stabilizers H_x for $x \in X$. The Haar measure μ is finite on compact subsets H_x, and when H is unimodular, $\mu(H_x)$ is constant on H-orbits, so we can define

(4) $$\mu(H\backslash\backslash X) := \sum_{x \in H\backslash X} \frac{1}{\mu(H_x)}.$$

Let $\Gamma \leq H$ be a discrete subgroup. Then the stabilizers Γ_x are compact and discrete, hence finite, and we can define

(5) $$Vol(\Gamma\backslash\backslash X) := \sum_{x \in \Gamma\backslash X} \frac{1}{|\Gamma_x|}.$$

(6) **Theorem ([BL]), 1.6.** *Let H be a locally compact group with Haar measure μ. Let X be a left H-set with compact open stabilizers and let $\Gamma \leq H$ be a discrete subgroup. The following conditions are equivalent.*

(1) $Vol(\Gamma\backslash\backslash X) < \infty$.

(2) Γ *is an H-lattice (hence H is unimodular), and*
$\mu(H\backslash\backslash X) < \infty$

In this case:
$$Vol(\Gamma\backslash\backslash X) = \mu(\Gamma\backslash H) \cdot \mu(H\backslash\backslash X)$$

3.2 Lattice Existence Theorem

(1) Theorem 3.1.6 suggests obvious necessary conditions for the existence of X-lattices on a locally finite tree X. Let $G = Aut(X)$ and let μ be a Haar measure on G. For G to contain X-lattices, it is necessary that :

(2) $\qquad\qquad\qquad$ (U) G is Unimodular

that is, μ is both left and right invariant, and

(3) $\qquad\qquad\qquad$ (FV) $\mu(G\backslash\backslash X) < \infty$

where the unimodularity of G allows us to define:

(4) $$\mu(G\backslash\backslash X) = \sum_{x \in V(G\backslash X)} \frac{1}{\mu(G_x)}.$$

(5) For G to contain *uniform lattices* we must strengthen (FV) to:

(F) $G\backslash X$ is finite.

We have the **Lattice Existence Theorem** ([BCR])

(6) $\qquad\qquad\qquad$ X admits a lattice $\iff$ (U) + (FV)

and

(7) $\qquad\qquad\qquad$ X admits a uniform lattice $\iff$ (U) + (F)

that is, the necessary conditions are also sufficient for the existence of an X-lattice.

We also have the:

(8) Uniform Existence Theorem ([BK]). Let X be a locally finite tree, and $G = Aut(X)$. The following conditions are equivalent:

(a) There exists a uniform X-lattice.

(b) We have:

(U) G is Unimodular, and

(F) $G \backslash X$ is finite.

(c) X is the universal cover of a finite connected graph.

(9) Under the conditions of the uniform existence theorem, we call X a *uniform tree*.

3.3 Existence of non-uniform lattices on uniform trees

We ask if a uniform tree always admits a non-uniform lattice. More generally, let X be a uniform tree, let $H \leq G = Aut(X)$ be a subgroup without inversions, and with quotient $p : X \longrightarrow H \backslash X$, and group of 'deck transformations':

$$G_H = \{g \in G \mid p \circ g = p\}.$$

We wish to know if there is a non-uniform lattice $\Gamma \leq G_H$. The following examples by [BL] indicate that this is not true in general:

(1) It may happen that $G = Aut(X)$ is discrete. In this case we say that X is *rigid*.

Example (Bass-Tits)

$A =$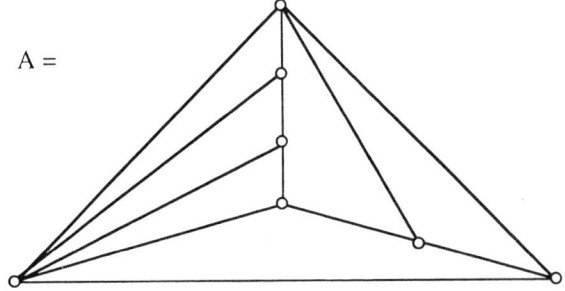

Let $X = \tilde{A}$. Then $G = Aut(X)$ is itself discrete ([BT]) and hence a uniform X-lattice so it cannot contain a non-uniform lattice.

(2) Let $X = \tilde{A}$, where

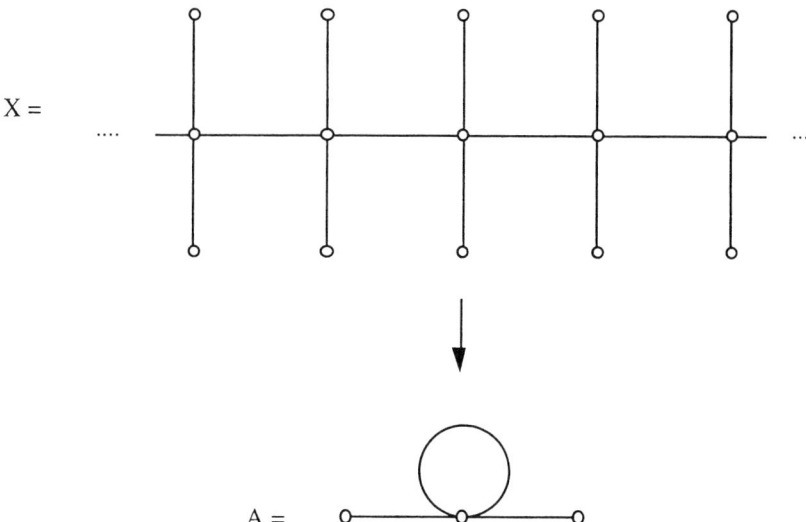

Then $G = Aut(X)$ is not discrete ([BL], [BK]), but all X-lattices are uniform. However, action of G on X is not *minimal*; that is, there is a proper G-invariant subtree, the central axis.

Ruling out these phenomena, we have the following, conjectured in an earlier version of ([BL]):

(3) **Conjecture (Existence of non-uniform lattices on uniform trees) ([BL]).** *Let X be a locally finite tree, and let $H \leq G = Aut(X)$ be a subgroup without inversions. Suppose that $\overline{H}$ is unimodular and $H \backslash X$ is finite (thus, X admits a uniform X-lattice). If G_H is not discrete and H (equivalently G_H) acts minimally on X, then there is a non-uniform X-lattice $\Gamma \leq G_H$.*

3.4 Existence of non-uniform coverings

We will give a constructive proof of conjecture 3.3.3 for the existence of non-uniform lattices on uniform trees. We shall construct our lattice Γ by constructing instead the appropriate $\Gamma\backslash\backslash X$.

(1) Given $H \leq G = Aut(X)$ and $G_H = \{g \in G \mid p \circ g = p\}$, satisfying:

(U) $\overline{H}$ is unimodular,

(F) $H\backslash X$ is finite,

(ND) G_H is not discrete,

(MIN) H acts minimally on X,

we build the quotient graph of groups $\mathbb{A} = H\backslash\backslash X$ on $A = H\backslash X$, and we form the corresponding indexed graph:

(2) $$(A, i) = I(H\backslash\backslash X) = I(G_H\backslash\backslash X).$$

(3) If there exists a non-uniform lattice $\Gamma \leq G_H$, then there exists an infinite (B, j), such that (B, j) is the edge-indexed graph for the graph of groups of the action of Γ on X, $(B, j) = I(\Gamma\backslash\backslash X)$, and we have a commutative diagram of natural projections:

(4)
$$\begin{array}{ccc} & X & \\ {}^{p_B}\swarrow & & \searrow{}^{p_A} \\ B & \xrightarrow{p} & A \end{array}$$

In fact

(5) $$p : (B, j) \longrightarrow (A, i)$$

will be a *covering* of edge indexed graphs. We recall from 2.2.4 that a covering of edge indexed graphs is a graph morphism $p : B \longrightarrow A$ with the property that the sum of the indices in the 'local fiber' above an edge equals the index on the edge below :

(6) $$i(e) = \sum_{f \in p_{(b)}^{-1}(e)} j(f).$$

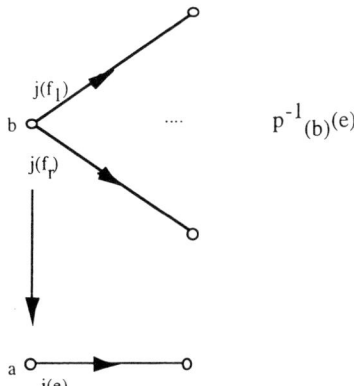

(7) Conversely, to construct Γ, it suffices to construct the covering:

$$p : (B, j) \longrightarrow (A, i)$$

such that (B, j) satisfies the conditions:

(INF) (B, j) is infinite,

(U) (B, j) is unimodular,

(FV) (B, j) has finite volume,

(BD) (B, j) has bounded denominators.

(8) We observe that (U) and (FV) are necessary for the existence of a lattice Γ (this follows immediately from the definitions). Conditions (U) and (BD) are nesessary and sufficient for the existence of a finite grouping $\mathbb{B}$ of (B, j) (see 1.4.9), and the fact that Γ is non-uniform is reflected in the condition that (B, j) is infinite.

(9) We can then choose a finite (faithful) grouping $\mathbb{B}$ of (B, j). The desired non-uniform X-lattice Γ is $\Gamma = \pi_1(\mathbb{B}, b)$ (see [B] or [S] for a precise definition of the fundamental group of a graph of groups). By 2.2.4, we have $\Gamma = \pi_1(\mathbb{B}, b)) \leq G_H$.

(10) We will obtain a combinatorial restatement of conjecture 3.3.3 once we translate our assumptions about (H, X) into combinatorial properties of the indexed graph: $(A, i) = I(H\backslash\backslash X)$.

(11) Condition (F), that $H\backslash X$ is finite, becomes:

(F) A is *finite*.

Condition (U), that $\overline{H}$ is unimodular, becomes:

(U) (A, i) is *unimodular*.

The condition (MIN), that H acts minimally on X, becomes:

(MIN) (A, i) has no *terminal vertices*:

$$deg_{(A,i)}(a) = \sum_{e \in E_0(a)} i(e) > 1 \text{ for all } a \in VA.$$

Condition (ND), that G_H is not discrete, translates into the combinatorial condition given by Bass and Tits in [BT]:

(NDR) (A, i) is *non-discretely ramified*:

ie. if (A, i) is minimal, there exists $e \in EA$ such that:

$$i(e) \geq 3 \text{ or } i(e) = 2 \text{ and } E_0(\partial_0 e) \neq \{e\}.$$

(12) We shall sometimes refer to a covering $p : (B, j) \longrightarrow (A, i)$ of edge-indexed graphs such that (B, j) has the properties (Inf), (U), (FV), (BD) as a *non-uniform covering* of the indexed quotient (A, i).

We give the combinatorial restatement of conjecture 3.3.3:

(13) Conjecture. *Let (A, i) be any connected, locally finite edge-indexed graph. Suppose that (A, i) satisfies (F), (U), (NDR) and (MIN). Then there is a covering $p : (B, j) \longrightarrow (A, i)$ of edge-indexed graphs such that (B, j) satisfies (U), (INF), (FV) and (BD).*

(14) We recall, from 0.18, that any connected locally finite graph can occur as the quotient of a tree by a lattice:

(15) Theorem ([BL], 4.17)('Any quotient theorem'). *Let A be any connected locally finite graph. Then there is a locally finite tree X and an X-lattice Γ such that $A \cong \Gamma \backslash X$.* □

4. Non-uniform coverings of indexed graphs with an arithmetic bridge

In this section, we prove the existence conjecture for non-uniform lattices on uniform trees (conjecture 3.3.3, and its combinatorial restatement 3.4.13) in a large number of cases; that is, when the indexed quotient graph contains an 'arithmetic bridge' (see definition 4.1.3).

4.1 Geometric and arithmetic bridges in indexed graphs

We introduce the notion of an 'arithmetic bridge' in an edge-indexed graph (A, i). This will be essential to the construction of non-uniform coverings of (A, i).

(1) Definition ((p,q)-geometric bridge).

Let A be a connected graph (A may be finite or infinite). We say that $\beta \subset EA$ is a **(p,q)-geometric bridge** *for A if:*

(i) $\beta \neq \varnothing$, β is oriented, $\beta \cap \overline{\beta} = \varnothing$,

(ii) $A \backslash (\beta \cup \overline{\beta})$ has $p + q$ connected components, $A_1, A_2, \ldots A_p, B_1, B_2, \ldots B_q$,

(iii) *for every $e \in \beta$ we have $\partial_0 e \in \mathcal{S}_\beta = A_1 \cup A_2 \cup \cdots \cup A_p$, the source of β and $\partial_1 e \in \mathcal{T}_\beta = B_1 \cup B_2 \cup \cdots \cup B_q$, the target of β,*

(iv) *the source $\mathcal{S}_\beta$ of β does not contain any target vertex, and the target $\mathcal{T}_\beta$ of β does not contain any source vertex; that is, for every $v \in A_1 \cup A_2 \cup \cdots \cup A_p$, $v \neq \partial_1 e$ for any $e \in \beta$, for every $v \in B_1 \cup B_2 \cup \cdots \cup B_q$, $v \neq \partial_0 e$ for any $e \in \beta$:*

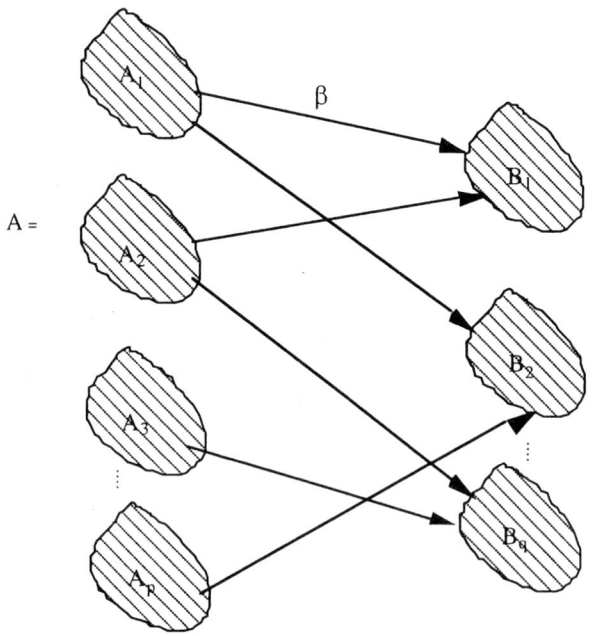

We say '$\partial_0 \beta = A_1 \cup A_2 \cup \cdots \cup A_p$' and '$\partial_1 \beta = B_1 \cup B_2 \cup \cdots \cup B_q$'.

(2) A $(1,1)$-geometric bridge β will be called a *geometric bridge*:

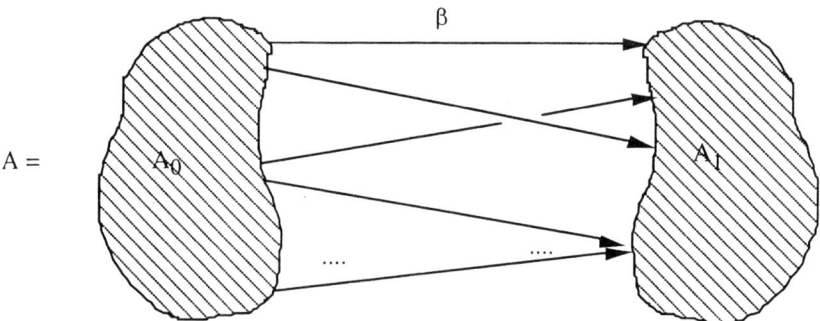

Let (A, i) be any (finite or infinite) connected edge-indexed graph.

(3) **Definition ((p,q)-arithmetic bridge).** *A (p,q)-geometric bridge β for A is called a **(p,q)-arithmetic bridge** for (A, i) if there exists a positive integer $d > 1$ such that $d \mid i(e)$ for every $e \in \beta$, say $i(e) = d i_0(e)$:*

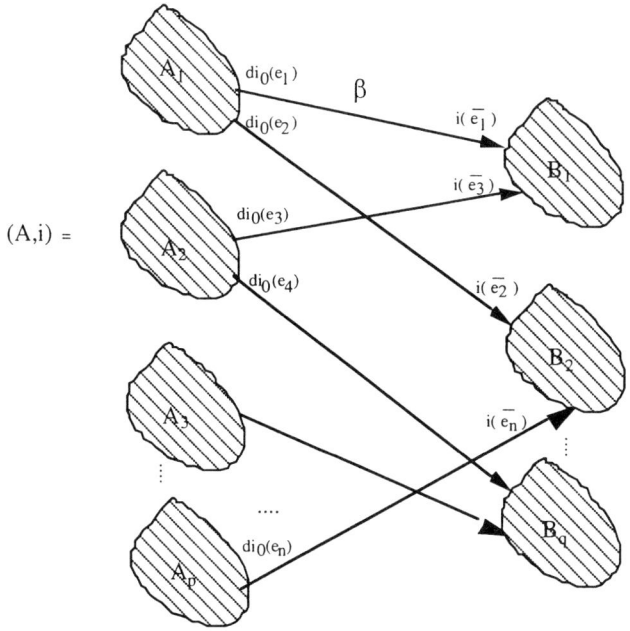

(4) We call d the *ramification factor* of β.

(5) A $(1,1)$-arithmetic bridge will be called an *arithmetic bridge*:

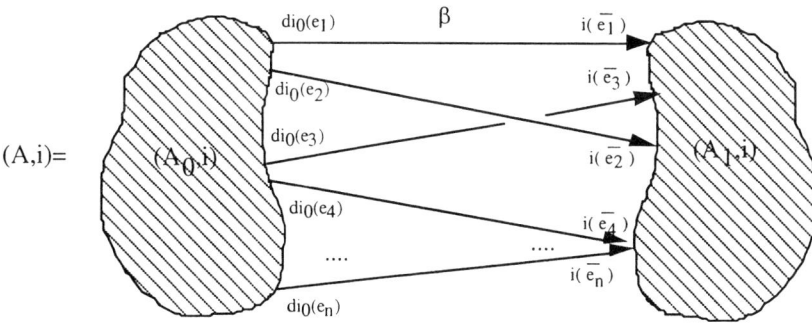

Our objective in this section is to prove the following:

(6) Theorem (Arithmetic bridge implies non-uniform covering). *Let (A,i) be an edge-indexed graph. Suppose that (A,i) satisfies (U), (F), (NDR), (MIN), and that*

(A, i) contains an arithmetic bridge β. Then (A, i) has a covering $p : (B, j) \longrightarrow (A, i)$ with the properties (U), (INF), (FV), (BD).

4.2 Changing the ramification factor of an arithmetic bridge

The following constructions will be needed to verify that non-uniform coverings of edge-indexed graphs with arithmetic bridges are unimodular.

(1) Construction (Changing the ramification factor of an arithmetic bridge).

if (A, i) is an indexed graph with arithmetic bridge β of ramification factor d, then we can make β an arithmetic bridge of ramification factor d', for any positive integer $d' > 0$, by replacing $di_0(e)$ by $d'i_0(e)$ for each positively oriented edge e of β. We write $(\frac{d'}{d})\beta$ for the new arithmetic bridge.

(2) Lemma (Changing the ramification factor is unimodular). *If (A, i) is a unimodular edge-indexed graph with arithmetic bridge β, then the indexed graph (A, i') obtained from (A, i) by replacing β by $(\frac{d'}{d})\beta$ is also unimodular.*

Proof. Let $E^+(\beta)$ be the set of positively oriented edges of β. For $e \in EA$ we define a new indexing $i'(e)$ as follows:

$$i'(e) = i(e), \text{ if } e \notin E^+(\beta),$$
$$d'i_0(e) = \frac{d'}{d}i(e), \text{ if } e \in E^+(\beta).$$

For $e \in EA$, set $\Delta(e) = \frac{i(\bar{e})}{i(e)}$, and $\Delta'(e) = \frac{i'(\bar{e})}{i'(e)}$.
Then for any $e \in EA$, we have:

$$\begin{aligned}
\Delta'(e) = \; & (\frac{d}{d'}) \cdot \Delta(e), \; e \in E^+(\beta), \\
& (\frac{d'}{d}) \cdot \Delta(e), \; \bar{e} \in E^+(\beta), \\
& \Delta(e), \; e \notin \beta.
\end{aligned}$$

Let γ be a (sufficiently long) closed path in A with initial (and hence terminal) vertex in the connected component A_0. Then γ crosses back and forth between A_0 and A_1, each time traversing an edge of β, returning finally to A_0.

It follows that γ traverses β and $\overline{\beta}$ an equal number of times, say r times. Thus:

$$\Delta'_{(A,i')}(\gamma) = (\frac{d}{d'})^r(\frac{d'}{d})^r\Delta(\gamma)$$
$$= \Delta_{(A,i)}(\gamma)$$
$$= 1, \text{ since } (A,i) \text{ is unimodular.}$$

$\square$

4.3 Gluing unimodular subgraphs along connected intersections

In this section, we describe a technique for 'gluing' together unimodular edge-indexed graphs in such a way that unimodularity is preserved.

(1) Lemma (Gluing unimodular subgraphs along connected intersection). *Let (A,i), (A_0,i), and (A_1,i) be indexed graphs such that $(A,i) = (A_0,i) \cap (A_1,i)$, and A_0, A_1 and $A_0 \cap A_1$ are connected. If (A_0,i) and (A_1,i) are unimodular, then (A,i) is unimodular.*

Proof. Observe that $(A_0 \cap A_1, i)$ is unimodular since it is a subgraph of a unimodular edge-indexed graph $((A_0,i)$ or $(A_1,i))$ (cf. 1.4.5).

Let γ be a (sufficiently long) closed path in $A = A_0 \cup A_1$ with initial (and hence terminal) vertex a_0 in $A_0 \cap A_1$. Then γ crosses back and forth between A_0 and A_1, say n times, each time passing through $A_0 \cap A_1$, returning finally to a_0 in $A_0 \cap A_1$.

Suppose that the j-th time γ passes through $A_0 \cap A_1$, γ passes through a vertex $a_j \in A_0 \cap A_1$ (there are n such vertices $a_1, \ldots, a_n$). For each $j = 1, 2, \ldots, n$, let γ_j be a closed path initiating at $a_j \in A_0 \cap A_1$, passing through $a_{j-1} \in A_0 \cap A_1$, and remaining entirely in $A_0 \cap A_1$. Since $(A_0 \cap A_1, i)$ is unimodular, we have $\Delta_{(A_0 \cap A_1, i)}(\gamma_j) = 1$.

For $t = 1, 2$, and $s = 0, 1, \ldots, n$, let $\gamma^{A_t}_{a_s a_{s+1}}$ denote the subpath of γ initiating at $a_s \in A_0 \cap A_1$, passing through A_t and terminating at $a_{s+1} \in A_0 \cap A_1$. For $t = 1, 2$, and $s = 0, 1, \ldots, n$, let $\gamma^j_{a_s a_{s+1}}$ denote the subpath of (the closed path) γ_j initiating at $a_s \in A_0 \cap A_1$ and terminating at $a_{s+1} \in A_0 \cap A_1$.

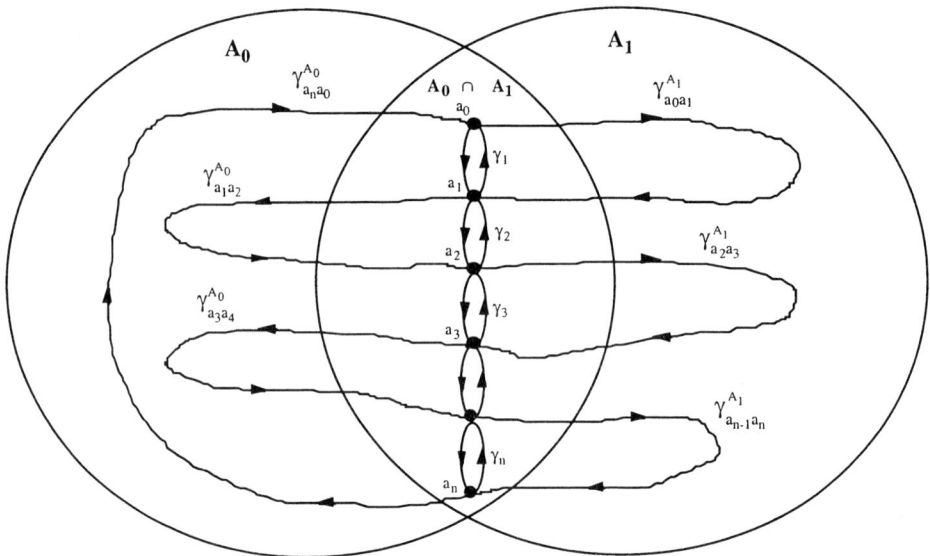

Consider the closed path γ' based at $a_0 \in A_0 \cap A_1$:

$$\gamma' = \gamma^{A_1}_{a_0 a_1} \cdot \gamma_1 \cdot \gamma^{A_0}_{a_1 a_2} \cdot \gamma_2 \cdot \gamma^{A_1}_{a_2 a_3} \cdot \gamma_3 \cdot \ldots \cdot \gamma^{A_1}_{a_{n-1} a_n} \cdot \gamma_n \cdot \gamma^{A_0}_{a_n q_0}.$$

Then $\Delta(\gamma') = \Delta(\gamma)$ since γ' is obtained from γ by inserting closed paths γ_j, $j = 1, 2, \ldots, n$ between a_s and a_{s+1}, $s = 0, 1, \ldots, n$ (and we have $\Delta_{(A_0 \cap A_1, i)}(\gamma_j) = 1$, $j = 1, 2, \ldots, n$).

Moreover, γ' can be expressed as a product of paths $\gamma' = \sigma_1 \sigma_2 \ldots \sigma_l$ such that σ_j is contained entirely in either A_0 or A_1; namely

$$\sigma_1 = \gamma^{A_1}_{a_0 a_1} \cdot \gamma^1_{a_1 a_0}$$

$$\sigma_2 = \gamma^1_{a_0 a_1} \cdot \gamma^{A_0}_{a_1 a_2} \cdot \gamma^2_{a_2 a_1} \cdot \gamma^1_{a_1 a_0}$$

$$\sigma_3 = \gamma^1_{a_0 a_1} \cdot \gamma^2_{a_1 a_2} \cdot \gamma^{A_1}_{a_2 a_3} \cdot \gamma^3_{a_3 a_2} \cdot \gamma^2_{a_2 a_1} \cdot \gamma^1_{a_1 a_0}$$

$$\vdots$$

Then each σ_{2j} is a closed path through A_0 based at $a_0 \in A_0 \cap A_1$, each σ_{2j+1} is a closed path through A_1 based at $a_0 \in A_0 \cap A_1$, and therefore:

$$\Delta_A(\gamma) = \Delta_A(\gamma')$$
$$= \Delta_{A_1}(\sigma_1)\Delta_{A_0}(\sigma_2)\ldots\Delta_{A_0}(\sigma_l)$$
$$= 1,$$

since (A_0, i) and (A_1, i) are unimodular. □

4.4 Open fanning of arithmetic bridges

In this section, we describe a technique to modify a unimodular edge indexed graph (A, i) with an arithmetic bridge β, in such a way that we preserve unimodularity, and obtain a new arithmetic bridge with a different ramification factor.

(1) Construction (Open fanning of arithmetic bridges I).

Suppose that (A, i) is an edge-indexed graph containing an arithmetic bridge β with ramification factor d. We represent (A, i) schematically as follows:

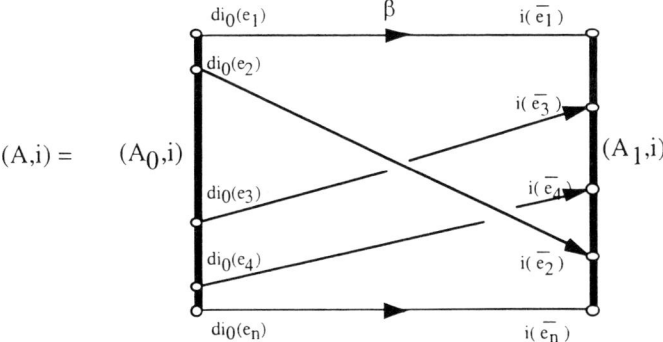

The *open fanning of β in (A, i)* is the edge-indexed graph (B, j) obtained by replacing β by d copies of $\frac{1}{d}\beta$; $\beta_1, \ldots \beta_d$, such that each positively oriented edge e of β_l in (B, j)

has index $i_0(e)$ for $l = 1, \ldots d$:

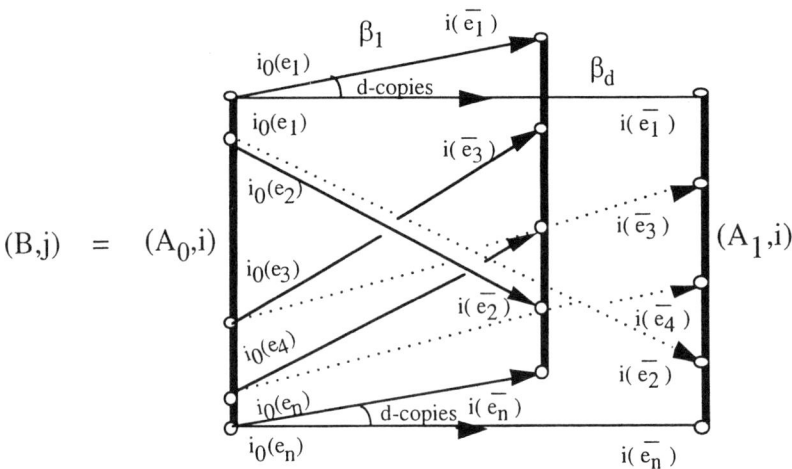

We observe that $p : (B, j) \longrightarrow (A, i)$ is a covering of indexed graphs.

(2) When β consists of a single (ramified) edge:

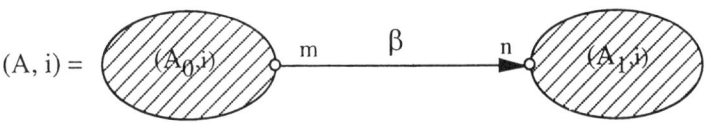

the open fanning of β in (A, i) coincides with the notion of 'open fanning of a separating edge' in [BL], (7.2). In this case, the edge β with its ramification index m is replaced by

m copies of $\frac{1}{m}\beta$, each with index 1:

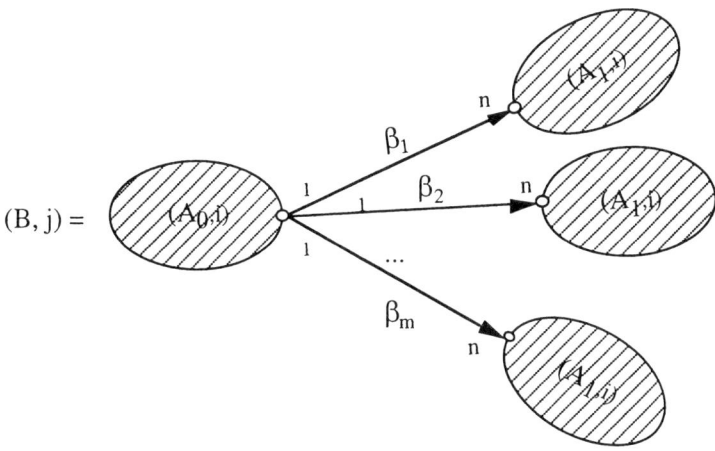

(3) Construction (Open fanning of arithmetic bridges II).

We shall also consider the following modification of open fanning:

Suppose that (A, i) is an edge-indexed graph containing an arithmetic bridge β:

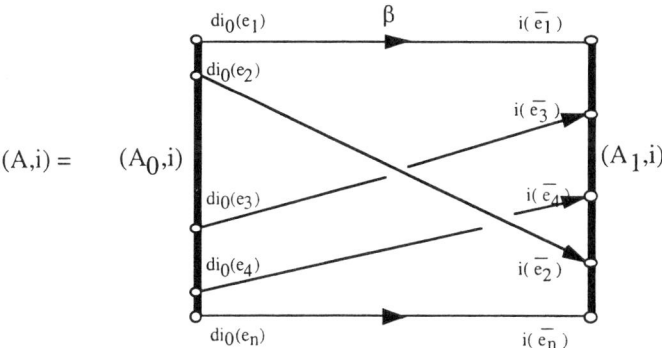

Rather than fanning open the arithmetic bridge β with its ramification factor d into d copies of $\frac{1}{d}\beta$, we obtain an indexed graph (B, j) by replacing β with $\beta^+ = \frac{1}{d}\beta$ and $\beta^- = \frac{d-1}{d}\beta$. Thus each edge e of β^+ has index $i_0(e)$, and each edge e of β^- has index

$(d-1)i_0(e)$:

$(B, j) =$

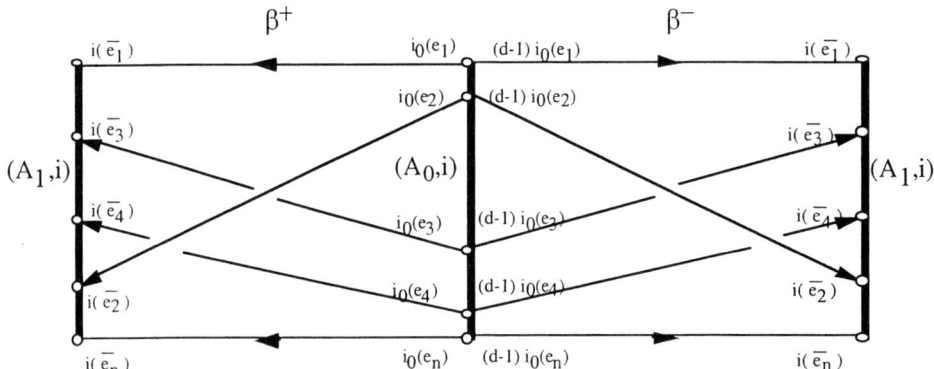

We observe that $p : (B, j) \longrightarrow (A, i)$ is a covering of indexed graphs.

The following lemma indicates that the process of open fanning preserves unimodularity:

(4) Lemma (Open fanning is unimodular). *Let (A, i) be a unimodular edge-indexed graph containing an arithmetic bridge β. Then the open fanning (I and II) of β in (A, i) is unimodular.*

Proof.

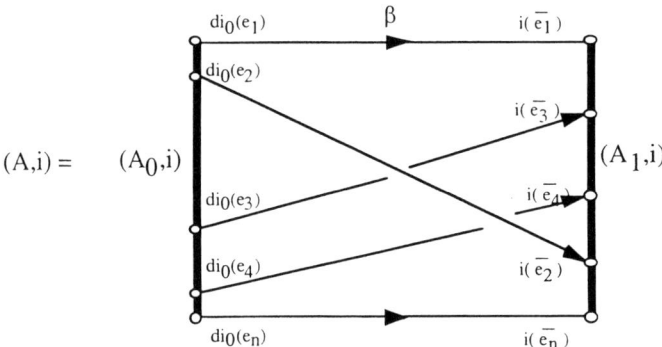

Let (A, i') and (A, i'') denote the indexed graphs obtained by changing the ramfication factor of β from d to 1, and from d to $d-1$ respectively. By lemma 4.2.2 (Changing the ramification factor is unimodular), (A, i') and (A, i'') are unimodular.

The open fanning (I) of β in (A,i) (see 4.4.1) is an indexed graph (B,j) obtained by 'gluing' d copies of (A,i') together along the connected subgraph A_0. Similarly, in open fanning (II) (see 4.4.3), we glue (A,i') and (A,i'') together along the connected subgraph A_0. By lemma 4.3.1 (Gluing unimodular subgraphs along connected intersection), the result of open fanning (I) (4.4.1) or open fanning (II) (4.4.3) is unimodular. $\square$

4.5 Indexed topological coverings

In this section, we describe a technique for constructing coverings of edge-indexed graphs by taking topological coverings of the underlying graph, and lifting the indexing in such a way that the projection is index preserving. The resulting *indexed topological covering* will automatically be unimodular, and an edge-indexed covering of the original indexed graph.

(1) Definition (Indexed topological coverings).

Let $p: B \longrightarrow A$ be a graph morphism. If $i: EA \longrightarrow \mathbb{Z}$ is an indexing on A, we can lift it to an indexing $j = i \circ p$ on B so that $p: (B,j) \longrightarrow (A,i)$ is index preserving: $i(p(f)) = j(f)$ for each $f \in B$. If (A,i) is unimodular, then so also is (B,j). In fact, if $\gamma = (e_1, \ldots, e_n)$ be a closed path in (B,j), then

$$\Delta_{(B,j)}(\gamma) = \Delta_{(A,i)}(p(\gamma)) \text{ since } p \text{ is index preserving.}$$

Since $p(\gamma)$ is closed and (A,i) is unimodular, we have $\Delta_{(A,i)}(p(\gamma)) = 1$, and so $\Delta_{(B,j)}(\gamma) = 1$.

We call a graph morphism $p: B \longrightarrow A$ a *topological covering* if the local map:

$$p_{(b)}: E_0(b) \longrightarrow E_0(p(b))$$

is bijective, for every $b \in VB$.

If (A,i) and (B,j) are indexed graphs, and $p: B \longrightarrow A$ is an index preserving topological covering, then $p: (B,j) \longrightarrow (A,i)$ is a covering of edge-indexed graphs.

We are now able to prove:

(2) Theorem (Arithmetic bridge implies non-uniform covering). *Let (A,i) be an edge-indexed graph. Suppose that (A,i) satisfies (U), (F), and that (A,i) contains an arithmetic bridge β. Then (A,i) has a covering $p : (B,j) \longrightarrow (A,i)$ with the properties (U), (INF), (FV), (BD).*

(3) Remark. *The existence of an arithmetic bridge $\beta \subset EA$ with $|\beta| \geq 2$ implies that (A,i) satisfies (NDR).*

Proof.

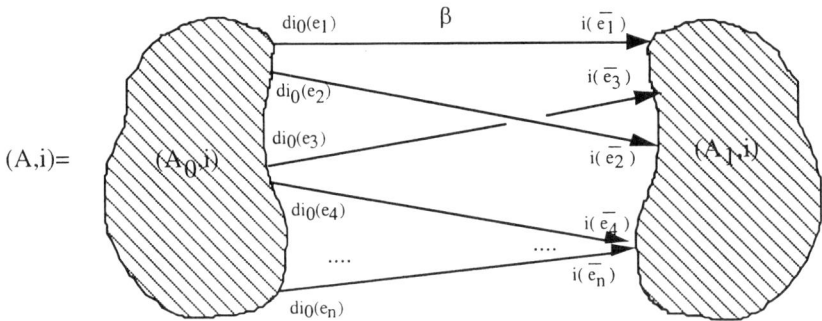

We assume that β has $n \geq 2$ edges. The case that β consists of a single edge will be treated in section 5.

4.6 Step 1 - Schematic diagram

We represent (A,i) schematically as follows, where we have choosen two edges e_1 and

e_n of β:

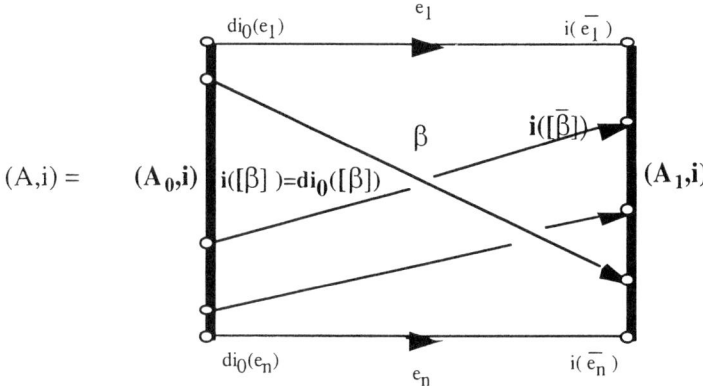

Let $[\beta] = \beta - \{e_1, e_n\}$. We have schematically denoted the indexing of $[\beta]$ as '$i([\beta]) = di_0([\beta])$'; more precisely, $i(e) = di_0(e)$ for every $e \in [\beta]$.

4.7 Step 2 - Construct topological covering

(1) We form a 3-fold topological covering $p: A_3 \to A$ and lift the indexing i_A to an indexing $(i \circ p)$ on A_3 such that p is index preserving. Then by 4.5.1, (A_3, i) is unimodular

and $p: (A_3, i) \to (A, i)$ is a covering of edge-indexed graphs.

$(A_3, i) =$

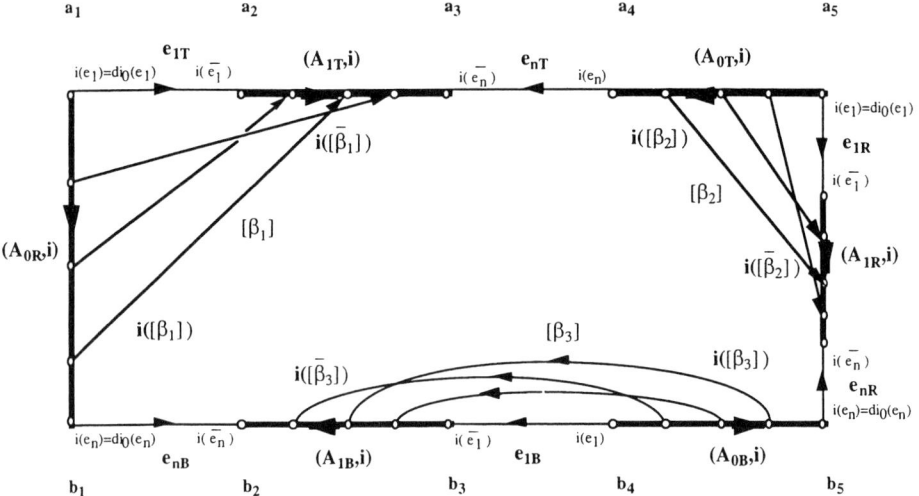

(2) We have denoted the three copies of β, as β_1, β_2, β_3 and the corresponding copies of A_0 and A_1 as A_{0R}, A_{OT}, A_{OB}, A_{1T}, A_{1L}, A_{1B} where the lower labels 'T', 'B', 'R', 'L'; 'top', 'bottom', 'right', 'left', respectively, signify the position in the schematic diagram of (A_3, i).

(3) The vertex and edge labels are suggested by the notation. We observe that β_1, β_2, and β_3 are each arithmetic bridges for (A_3, i).

(4) Observe that the paths from a_1 to a_5 and b_1 to b_5, are both liftings of closed paths in (A, i), and since p is index-preserving:

$$\frac{\Delta a_5}{\Delta a_1} = \frac{\Delta b_5}{\Delta b_1} = 1$$

since (A, i) is unimodular.

4.8 Step 3 - Change the ramification factor

(1) We form a new indexed graph $(R_0^{(0)}, i)$ from (A_3, i) as follows: we open fan β_1 to '$\frac{d-1}{d}\beta_1$' and '$\frac{1}{d}\beta_1$' (that is, we apply open fanning II (see 4.4.3)) to β_1 which is an arithmetic bridge in (A_3, i) from A_{OR} to $A_3 - \{\beta_1 \cup A_{1T}\}$) :

$(R_0^{(0)}, i) =$

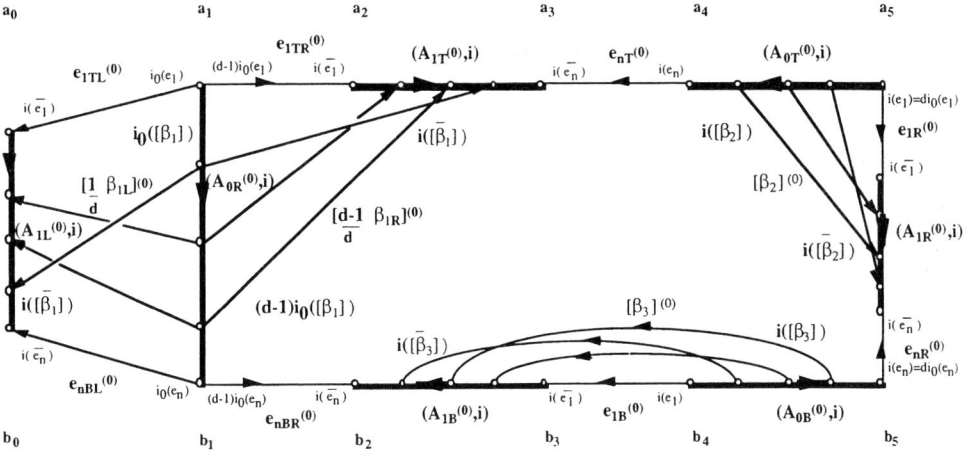

(2) By lemma 4.4.4 (Open fanning is unimodular), $(R_0^{(0)}, i)$ is unimodular. We observe that:

(3) $$\frac{\Delta a_5}{\Delta a_1} = \frac{i(\overline{e_1})}{(d-1)i_0(e_1)} \cdot \frac{\Delta_{(A_1,i)}a_3}{\Delta_{(A_1,i)}a_2} \cdot \frac{di_0(e_n)}{i(\overline{e_n})} \cdot \frac{\Delta_{(A_0,i)}a_5}{\Delta_{(A_0,i)}a_4} = \frac{d}{d-1}$$

by unimodularity of (A, i). Similarly,

(4) $$\frac{\Delta b_5}{\Delta b_1} = \frac{d}{d-1}.$$

(5) Moreover, the projection $p \colon (R_0^{(0)}, i) \to (A, i)$ is a covering of edge-indexed graphs.

(6) We observe that $[\beta_2] \cup e_{1R}^{(0)} \cup e_{nR}^{(0)}$ is an arithmetic bridge in $(R_0^{(0)}, i)$ from

$$R_0^{(0)} - \{[\beta_2] \cup e_{1R}^{(0)} \cup e_{nR}^{(0)} \cup A_{1R}^{(0)}\}$$

to $A_{1R}^{(0)}$.

(7) Next, we form a new indexed graph $(R^{(0)}, i)$, from $(R_0^{(0)}, i)$, by changing the ramification factor of the arithmetic bridge $[\beta_2] \cup e_{1R}^{(0)} \cup e_{nR}^{(0)}$ in $(R_0^{(0)}, i)$ from d to 1 (using 4.2.1):

$(R^{(0)}, i) =$

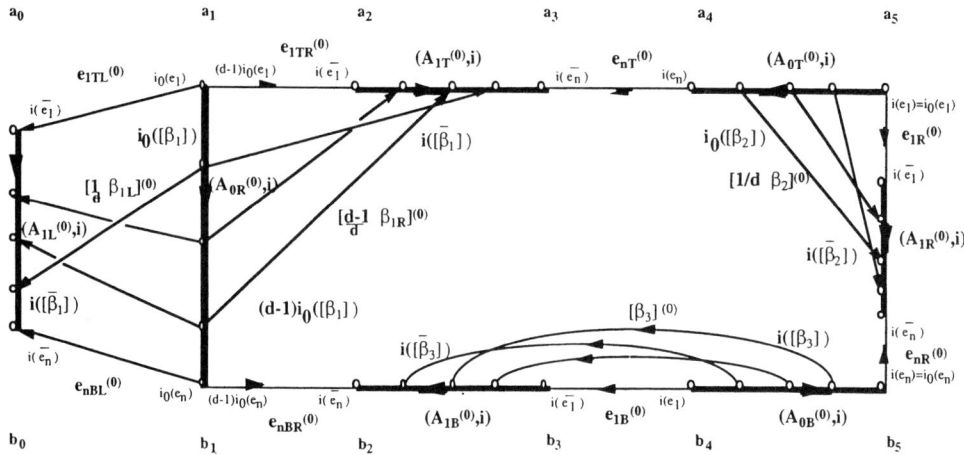

and by lemma 4.2.2 (Changing the ramification factor is unimodular), $(R^{(0)}, i)$ is unimodular.

(8) The notation $[\frac{1}{d}\beta_{1L}]^{(0)}$ denotes $[\beta_1]$ with its ramification factor changed from d to 1, $\left[\frac{d-1}{d}\beta_{1R}\right]^{(0)}$ denotes $[\beta_1]$ with its ramification-factor changed from d to d-1.

(9) The upper label (0) signifies that $[\frac{1}{d}\beta_{1L}]$ and $[\frac{d-1}{d}\beta_{1R}]$ belong to $R^{(0)}$; the lower labels 'L' and 'R' denote 'left' and 'right' respectively.

4.9 Step 4 - Construct rectangles

For $k = 1, 2, 3, \ldots$ let $(R^{(k)}, i)$ be the following edge-indexed graph:

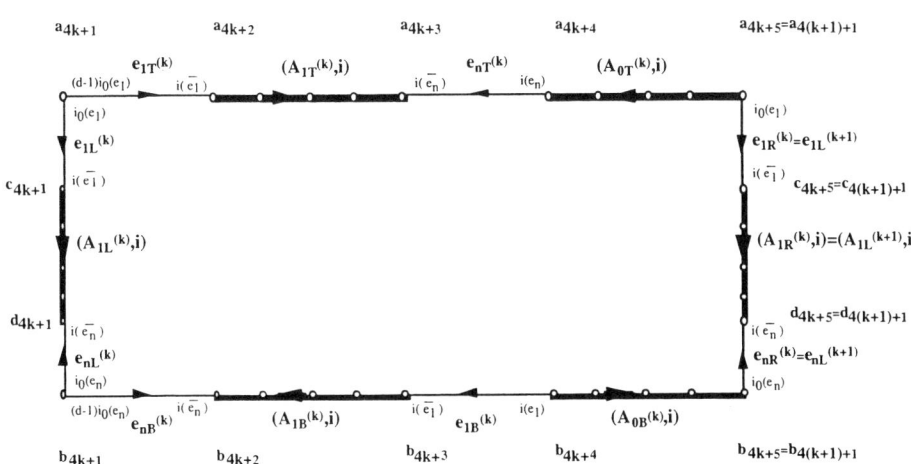

where the lower labels 'T', 'B', 'L', 'R' on edges $e_1^{(k)}$, $e_n^{(k)}$ and graphs $A_1^{(k)}$, $A_0^{(k)}$ indicate 'top', 'bottom', 'left' and 'right' respectively and signify the position within $R^{(k)}$.

(1) For each $k = 1, 2, 3, \ldots$ the 'rectangle' $(R^{(k)}, i)$ has as its 'top' a path from a_{4k+1} to a_{4k+5} which coincides with the path from a_1 to a_5 in $(R^{(0)}, i)$ (as described in 4.8), a path from b_{4k+1} to b_{4k+5} which coincides with the path from b_1 to b_5 in $(R^{(0)}, i)$, and paths from a_{4k+1} to b_{4k+1} and a_{4k+5} to b_{4k+5} each of which coincide with the path from a_5 to b_5 in $(R^{(0)}, i)$. Therefore,

(2)
$$\frac{\Delta a_{4k+5}}{\Delta a_{4k+1}} = \frac{\Delta b_{4k+5}}{\Delta b_{4k+1}} = \frac{d}{d-1}$$
$$\frac{\Delta b_{4k+1}}{\Delta a_{4k+1}} = \frac{\Delta b_{4k+5}}{\Delta a_{4k+5}},$$

and it follows easily that $(R^{(k)}, i)$ is unimodular.

4.10 Step 5 - Glue rectangles iteratively

We construct an infinite indexed graph from $(R^{(0)}, i)$ and $(R^{(k)}, i)$, $k = 1, 2, 3, \ldots$ by an infinite sequence of gluings: we identify the edges $e_{1R}^{(0)}$ and $e_{nR}^{(0)}$ and the subgraph $A_{1R}^{(0)}$ of $(R^{(0)}, i)$ with $e_{1L}^{(1)}$ and $e_{nL}^{(1)}$ and $A_{1L}^{(1)}$ of $(R^{(1)}, i)$, respectively.

(1) For $k = 1, 2, 3, \ldots$ we identify $e_{1R}^{(k)}$ and $e_{nR}^{(k)}$ and $A_{1R}^{(k)}$ of $(R^{(k)}, i)$ with $e_{1L}^{(k+1)}, e_{nL}^{(k+1)}$ and $A_{1L}^{(k+1)}$ of $(R^{(k+1)}, i)$, respectively.

(2) We denote the resulting indexed graph by $(R^{(\infty)}, i)$, which is shown in fig 4.10.2.

(3) By 4.8 (step 3), the indexed graph $(R^{(0)}, i)$ is unimodular, and by 4.9 (step 4), $(R^{(k)}, i)$ is unimodular, for $k = 1, 2, 3, \ldots$. Moreover, we have glued $(R^{(0)}, i)$ to $(R^{(1)}, i)$ and $(R^{(k)}, i)$ to $(R^{(k+1)}, i)$ respectively, for $k = 1, 2, 3, \ldots$ along connected subgraphs:

$$\{e_{1R}^{(k)} = e_{1L}^{(k+1)}\} \cup \{A_{1R}^{(k)} = A_{1L}^{(k+1)}\} \cup \{e_{nR}^{(k)} = e_{nL}^{(k+1)}\}.$$

We apply lemma 4.3.1 (Gluing unimodular subgraphs along connected intersection) to verify that the indexed graph $(R^{(\infty)}, i)$ is unimodular.

(4) We compute in $(R^{(\infty)}, i)$ for each $s = 1, 2, 3, \ldots$

$$\frac{\Delta a_{4s+1}}{\Delta a_1} = \left(\frac{d}{d-1}\right)^s = \frac{\Delta b_{4s+1}}{\Delta b_1},$$

$$\frac{\Delta c_{4s+1}}{\Delta a_1} = \left(\frac{d}{d-1}\right)^s \frac{i(\overline{e_1})}{i_0(e_1)},$$

$$\frac{\Delta d_{4s+1}}{\Delta b_1} = \left(\frac{d}{d-1}\right)^s \frac{i(\overline{e_n})}{i_0(e_n)}.$$

Since $d > 1$, it follows easily that $(R^{(\infty)}, i)$ has finite volume. We observe, however, that $(R^{(\infty)}, i)$ does *not* have bounded denominators.

□

(5) **Lemma (Adjoining an edge).** *let (A, i) and (A_0, i) be indexed graphs such that (A, i) is obtained from (A_0, i) by attaching an edge e to vertices $a, b \in VA_0$. If (A_0, i) is unimodular and*

$$\frac{\Delta_{A_0} b}{\Delta_{A_0} a} = \frac{i(\overline{e})}{i(e)}$$

then (A, i) is unimodular.

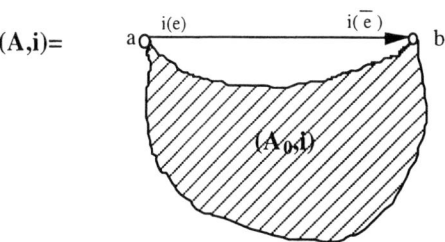

Proof. Obvious. □

(6) We fix the following notation: for $k = 1, 2, 3, \ldots$

$$[\frac{d-1}{d}\beta_{2T}]^{(k)} \text{ denotes } [\beta_2] \text{ with its ramification}$$
factor changed from d to d-1,

$$[\frac{1}{d}\beta_{2R}]^{(k)} \text{ denotes } [\beta_2] \text{ with its ramification}$$
factor changed from d to 1,

$$[\beta_3]^{(k)} \text{ denotes } [\beta_3].$$

The upper labels (k) indicate the k-th rectangle. The lower labels 'T', 'R' denote 'top' and 'right'.

4.11 Step 6 - Adjoin bridges

(1) We construct an indexed graph (B^-, j^-) from $(R^{(\infty)}, i)$ as follows: for $k = 1, 2, 3, \ldots$ we adjoin:

$$[\frac{1}{d}\beta_{2R}]^{(k)} \text{ to } (R^{(\infty)}, i) \text{ from } A^{(k)}_{OT} \text{ to} A^{(k)}_{1R},$$
$$[\frac{d-1}{d}\beta_{2T}]^{(k)} \text{ to } (R^{(\infty)}, i) \text{ from } A^{(k)}_{OT} \text{ to} A^{(k+1)}_{1T},$$
$$[\beta_3]^{(k)} \text{ to } (R^{(\infty)}, i) \text{ from } A^{(k)}_{OB} \text{ to } A^{(k)}_{1B}.$$

(2) The resulting edge indexed graph (B^-, j^-) is depicted in fig 4.11.2.

(3) Proposition. *The indexed graph (B^-, j^-) is unimodular and has finite volume.*

Proof. By 4.10 (step 5), the indexed graph $(R^{(\infty)}, i)$ is unimodular. For $k = 1, 2, 3, \ldots$, Let $e \in [\frac{1}{d}\beta_{2R}]^{(k)}$ we attach

$$\partial_0 e \text{ to } (R^{(\infty)}, i) \text{ at } a \in A_{OT}^{(k)},$$

$$\partial_1 e \text{ to } (R^{(\infty)}, i) \text{ at } b \in A_{1R}^{(k)}.$$

Let '$\frac{1}{d}(\frac{\Delta_A b}{\Delta_A a})$' denote $\frac{\Delta_A b}{\Delta_A a}$ with all occurences of d changed to 1. Then

$$\frac{\Delta_{R^{(\infty)}} b}{\Delta_{R^{(\infty)}} a} = \frac{1}{d}\left(\frac{\Delta_A b}{\Delta_A a}\right)$$

$$=_A \frac{d}{1}\left(\frac{i(\bar{e})}{d_{i_0}(e)}\right)$$

$$=_A \frac{i(\bar{e})}{i_0(e)}$$

$$=_{R^{(\infty)}} \frac{j^-(\bar{e})}{j^-(e)},$$

and thus by lemma 4.10.5 (Adjoining an edge), the result of adjoining $[\frac{1}{d}\beta_{2R}]^{(k)}$ is unimodular.

Let $e \in [\frac{d-1}{d}\beta_{2T}]^{(k)}$ we attach

$$\partial_0 e \text{ to } (R^{(\infty)}, i) \text{ at } a \in A_{OT}^{(k)},$$

$$\partial_1 e \text{ to } (R^{(\infty)}, i) \text{ at } b \in A_{1T}^{(k)}.$$

Let '$\frac{d-1}{d}(\frac{\Delta_A b}{\Delta_A a})$' denote $\frac{\Delta_A b}{\Delta_A a}$ with all occurences of d changed to d-1. Then

$$\frac{\Delta_{R^{(\infty)}} b}{\Delta_{R^{(\infty)}} a} = \frac{d-1}{d}\left(\frac{\Delta_A b}{\Delta_A a}\right)$$

$$=_A \frac{d}{d-1}\left(\frac{i(\bar{e})}{d_{i_0}(e)}\right)$$

$$=_A \frac{i(\bar{e})}{(d-1)i_0(e)}$$

$$=_{R^{(\infty)}} \frac{j^-(\bar{e})}{j^-(e)}.$$

By lemma 4.10.5 (Adjoining an edge), the result of adjoining $[\frac{d-1}{d}\beta_{2T}]^{(k)}$ is unimodular.

Let $e \in [\beta_3]^{(k)}$ we attach

$$\partial_0 e \text{ to } a \in A_{0B}^{(k)},$$
$$\partial_1 e \text{ to } b \in A_{1B}^{(k)}.$$

Then

$$\frac{\Delta_{R^{(\infty)}} b}{\Delta_{R^{(\infty)}} a} = \frac{\Delta_A b}{\Delta_A a} =_A \frac{i(\bar{e})}{i(e)} =_{R^{(\infty)}} \frac{j^-(\bar{e})}{j^-(e)}$$

and by lemma 4.10.5 (Adjoining an edge), the result of adjoining $[\beta_3]^{(k)}$ is unimodular.

Since the result of adjoining $[\frac{1}{d}\beta_{2R}]^{(k)}$, $[\frac{d-1}{d}\beta_{2T}]^{(k)}$, $[\beta_3]^{(k)}$ is unimodular for $k = 1, 2, 3, \ldots$ it follows that the resulting indexed graph (B^-, j^-) inherits finite volume from $(R^{(\infty)}, i)$. □

(4) Remark. *The indexed graph (B^-, j^-) does not have bounded denominators.*

(5) Proposition. *There is a morphism $q\colon (B^-, j^-) \to (A, i)$ which is a covering of indexed graphs.*

Proof. Recall that for the arithmetic bridge $\beta \subset EA$, $[\beta]$ denotes $\beta - \{e_1, e_n\} \subset EA$. For $j = 1, 2, 3$ and $k = 1, 2, \ldots [\beta_j]^{(k)} \subset EB^-$ denotes the j-th copy of $[\beta]$ in $R^{(k)}$.

We define a morphism $q\colon B^- \to A$:

$$q|_{A_{1R}^{(0)}} \colon A_{1R}^{(0)} \xrightarrow{\cong} A_1$$
$$q(e_{1_T L}^{(0)}) = e_1$$
$$q|_{[\beta_{1L}]^{(0)}} \colon [\beta_{1L}]^{(0)} \xrightarrow{\cong} [\beta]$$
$$q(e_{n_B L}^{(0)}) = e_n$$
$$q(e_{1_T R}^{(0)}) = e_1$$
$$q|_{[\beta_{1R}]^{(0)}} \colon [\beta_{1R}]^{(0)} \xrightarrow{\cong} [\beta]$$
$$q(e_{n_b R}^{(0)}) = e_n$$

for $k = 1, 2, 3, \ldots$

$$q\mid_{A_{1T}^{(k)}} : A_{1T}^{(k)} \xrightarrow{\cong} A_1$$

$$q\mid_{A_{0T}^{(k)}} : A_{0T}^{(k)} \xrightarrow{\cong} A_0$$

$$q\mid_{A_{0B}^{(k)}} : A_{0B}^{(k)} \xrightarrow{\cong} A_1$$

$$q\mid_{A_{1B}^{(k)}} : A_{1B}^{(k)} \xrightarrow{\cong} A_1$$

$$q\mid_{A_{1L}^{(k)}} : A_{1L}^{(k)} \xrightarrow{\cong} A_1$$

$$q(e_{nT}^{(k)}) = e_n$$

$$q(e_{1B}^{(k)}) = e_1$$

$$q(e_{1T}^{(k+1)}) = e_1$$

$$q(e_{nB}^{(k+1)}) = e_n$$

$$q(e_{1L}^{(k+1)}) = e_1$$

$$q(e_{nL}^{(k+1)}) = e_n$$

$$q\mid_{\frac{1}{d}[\beta_{2R}]^{(k)}} : [\frac{1}{d}\beta_{2R}]^{(k)} \xrightarrow{\cong} [\beta]$$

$$q\mid_{\frac{(d-1)}{d}[\beta_{2T}]^{(k)}} : [\frac{(d-1)}{d}\beta_{2T}]^{(k)} \xrightarrow{\cong} [\beta]$$

$$q\mid_{[\beta_3]^{(k)}} : [\beta_3]^{(k)} \xrightarrow{\cong} [\beta]$$

The projection $q: (B^-, j^-) \to (A, i)$ 'erases upper and lower indices'.

Let $e \in EA_0$ with $q^{-1}(\partial_0 e) = b \in VB^-$. Then $q_{(b)}^{-1}(e)$ is isomorphic to e, so

$$\sum_{f \in q_{(b)}^{-1}(e)} j^-(f) = i(e).$$

For $e \in EA_1$, with $q^{-1}(\partial_0 e) = b \in VB^-$, $q_{(b)}^{-1}(e)$ is isomorphic to e, so

$$\sum_{f \in q_{(b)}^{-1}(e)} j^-(f) = i(e).$$

Now let $e \in \beta$, with $q^{-1}(\partial_0 \bar{e}) = b \in VB^-$, then $q_{(b)}^{-1}(\bar{e})$ is isomorphic to $\bar{e}$ so

$$\sum_{f \in q_{(b)}^{-1}(\bar{e})} j^-(f) = i(\bar{e}).$$

In order to compute the local fibers over each $e \in \beta$, we describe a labeling scheme for the edges of $\beta_{1L}^{(0)}$, $\beta_{1R}^{(0)}$, $\beta_{2R}^{(k)}$, $\beta_{2T}^{(k)}$ and $\beta_3^{(k)}$ in B^-, for each $k = 0, 1, 2, \ldots$

A typical edge is denoted $_{(j)}e_{lP}^{(k)}$, where $j = 1, 2$ or 3 denotes the j-th copy of β, $k = 1, 2, 3, \ldots$ denotes the 'kth-rectangle' of B^-, $l = 1, 2, \ldots, n$ denotes the l-th edge of β and P denotes 'position'; that is, L, R, T, B; left, right, top, or bottom, respectively.

For $e \in \beta$, we have local pictures :

(I) The local fiber in $R^{(0)}$ above β_1.

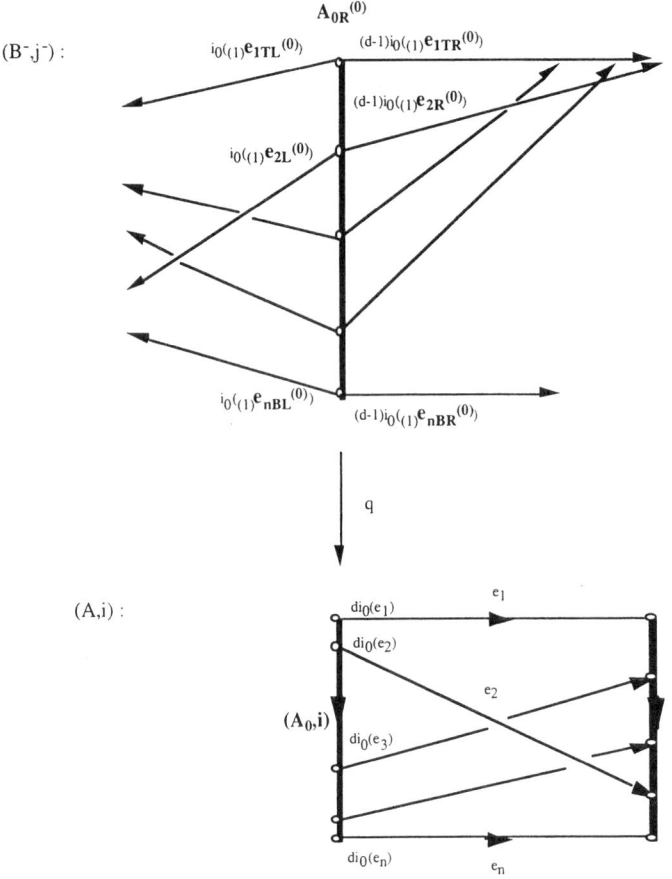

For $j = 1, \ldots, n$ let $b_j^{(0)} = \partial_0({}_{(1)}e_{jR}^{(0)}) = \partial_0({}_{(1)}e_{jL}^{(0)})$.

Then for $j = 1, \ldots, n$,

$$q({}_{(1)}e_{jR}^{(0)}) = e_j$$
$$q({}_{(1)}e_{jL}^{(0)}) = e_j$$
$$q(\partial_0 b_j^{(0)}) = \partial_0 e_j.$$

For each $e_j \in \beta$, $j = 1, \ldots, n$

$$\sum_{f \in q^{-1}_{(b_j^{(0)})}(e_j)} j^-(f) = (d-1)i_0(_{(1)}e_{jR}^{(0)}) + i_0(_{(1)}e_{jL}^{(0)})$$

$$= (d-1)i_0(e_j) + i_0(e_j)$$

$$= di_0(e_j)$$

$$= i(e_j).$$

52   LISA CARBONE

(II) The local fiber in $R^{(k)}$ above β_2, $k = 0, 1, 2 \ldots$.

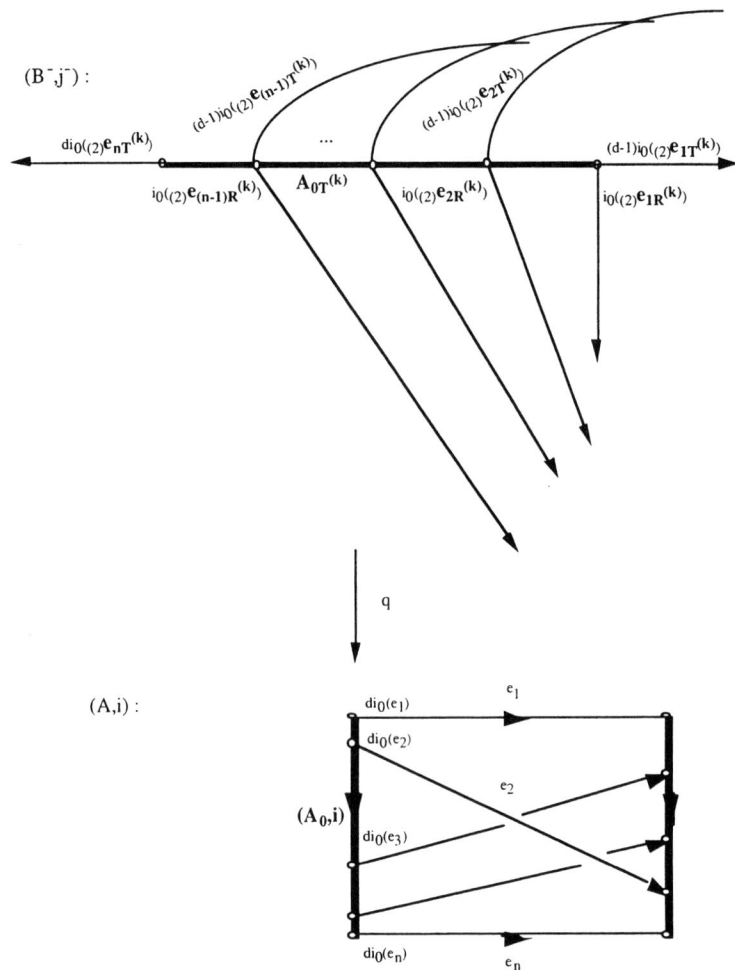

For $j = 1, \ldots, n$ let

$$b_j^{(k)} = \partial_0\big(_{(2)}e_{jR}^{(k)}\big) = \partial_0\big(_{(2)}e_{jT}^{(k)}\big)$$

For $k = 0, 1, 2 \ldots$, $j = 1, 2, \ldots, n-1$

$$q(_{(2)}e_{jR}^{(k)}) = e_j$$
$$q(_{(2)}e_{jT}^{(k)}) = e_j$$
$$q(_{(2)}e_{nT}^{(k)}) = e_n$$
$$q(b_j^{(k)}) = \partial_0 e_j$$

For $e_j \in \beta$, $j = 1, 2, \ldots, n-1$, $k = 0, 1, 2 \ldots$

$$\sum_{f \in q^{-1}_{(b_j^{(k)})}(e_j)} j^-(f) = (d-1)i_0(_{(2)}e_{jT}^{(k)}) + i_0(_{(2)}e_{jR}^{(k)})$$

$$= (d-1)i_0(e_j) + i_0(e_j)$$

$$= di_0(e_j)$$

$$= i(e_j).$$

For $e_n \in \beta$:

$$\sum_{f \in q^{-1}_{(b_n^{(k)})}(e_n)} j^-(f) = di_0(_{(2)}e_{nT}^{(k)})$$

$$= i_0(e_n).$$

(III) The local fiber in $R^{(k)}$ above β_3, $k = 0, 1, 2 \ldots$.

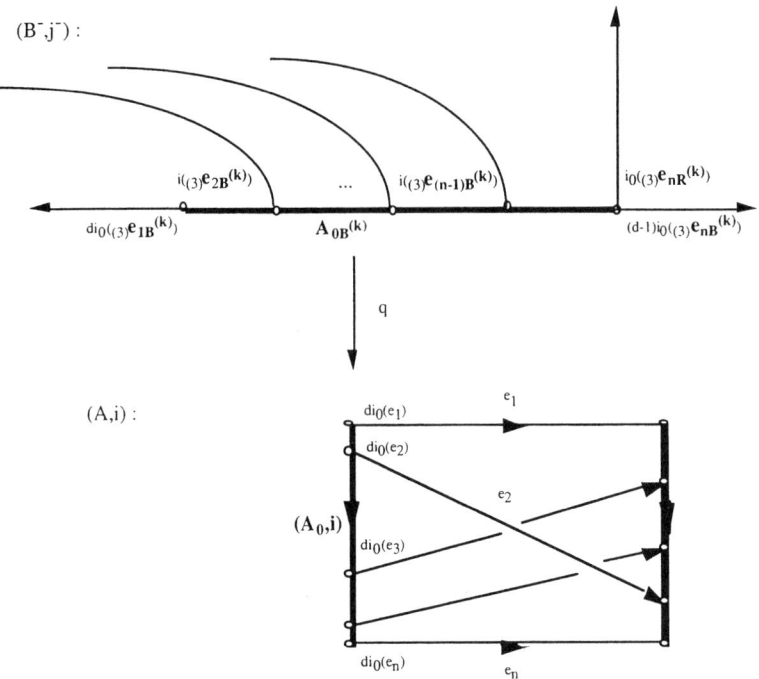

For $j = 1, \ldots, n-1$ let

$$b_j^{(k)} = \partial_0({}_{(3)}e_{jB}^{(k)})$$
$$b_n^{(k)} = \partial_0({}_{(3)}e_{nR}^{(k)}) = \partial_0({}_{(3)}e_{nB}^{(k+1)})$$

Then for $k = 0, 1, 2 \ldots$, $j = 1, 2, \ldots, n-1$

$$q({}_{(3)}e_{jB}^{(k)}) = e_j$$
$$q({}_{(3)}e_{nB}^{(k+1)}) = e_n$$
$$q({}_{(3)}e_{nR}^{(k)}) = e_n$$
$$q(b_j^{(k)}) = \partial_0 e_j, \quad j = 1, \ldots, n.$$

For $e_j \in \beta$, $j = 1, \ldots, n-1$, $k = 0, 1, 2 \ldots$

$$\sum_{f \in q_{b_j^{(k)}}^{-1}(e_j)} j^-(f) = i(e_j).$$

For $e_n \in \beta$

$$\sum_{f \in q_{b_n^{(k)}}^{-1}(e_n)} j^-(f) = (d-1)i_0(_{(3)}e_{nB}^{(k+1)}) + i_0(_{(3)}e_{nR}^{(k)})$$

$$= di_0(e_n)$$

$$= i(e_n). \square$$

4.12 Step 7 - Multiple open fanning

(1) We recall that the indexed graph (B^-, j^-) constructed in 4.11 (step 6) has the properties:

(INF) B^- is infinite,

(U) (B^-, j^-) is unimodular,

(FV) (B^-, j^-) has finite volume.

and that the projection $q: (B^-, j^-) \to (A, i)$ is a covering of edge-indexed graphs.

(2) We recall, moreover, that (B^-, j^-) does *not* have bounded denominators.

(3) We construct from (B^-, j^-) an indexed graph (B, j) with the properties that (INF), (U), (FV) such that (B, j) also has bounded denominators (BD) and such that $p: (B, j) \to (A, i)$ is a covering of edge-indexed graphs.

We refer to fig 4.11.2 (B^-, j^-) and observe that for each $k = 1, 2, 3, \ldots$ the set:

(4) $$\beta^{(k)} = e_{1T}^{(k)} \cup [\frac{d-1}{d}\beta_{2T}]^{(k-1)} \cup e_{nB}^{(k)} \subset EB^-$$

is an arithmetic bridge for (B^-, j^-) of ramification factor $(d-1)$.

(5) We obtain a graph (B, j) from (B^-, j^-) by '*multiple open fanning*' $\beta^{(k)}$, for each $k = 1, 2, 3, \ldots$. That is, for each $k = 1, 2, 3, \ldots$ we replace the arithmetic bridge $\beta^{(k)}$

with its ramification factor $(d-1)$ by $(d-1)$ copies of '$\frac{1}{d-1}\beta^{(k)}$' with ramification factor 1. More precisely, each rectangle $(R^{(k)}, i)$ of (B^-, j^-):

$(R^{(k)}, i) =$

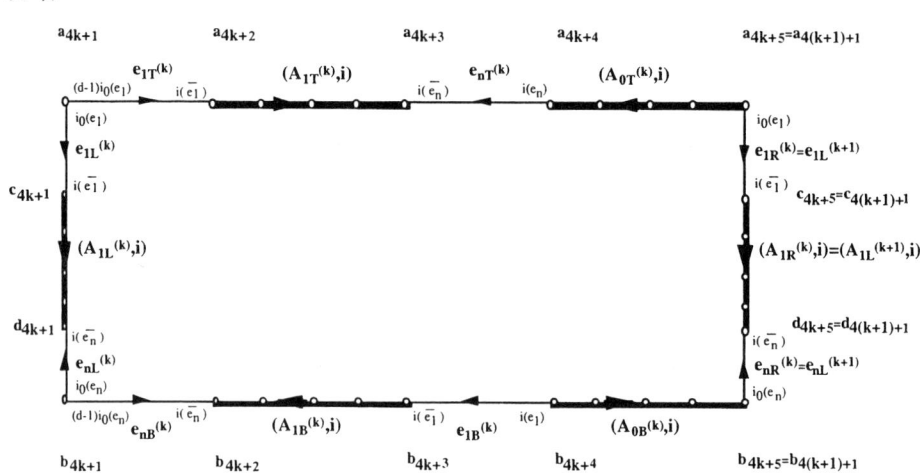

is replaced by $(d-1)$ rectangles of the form:

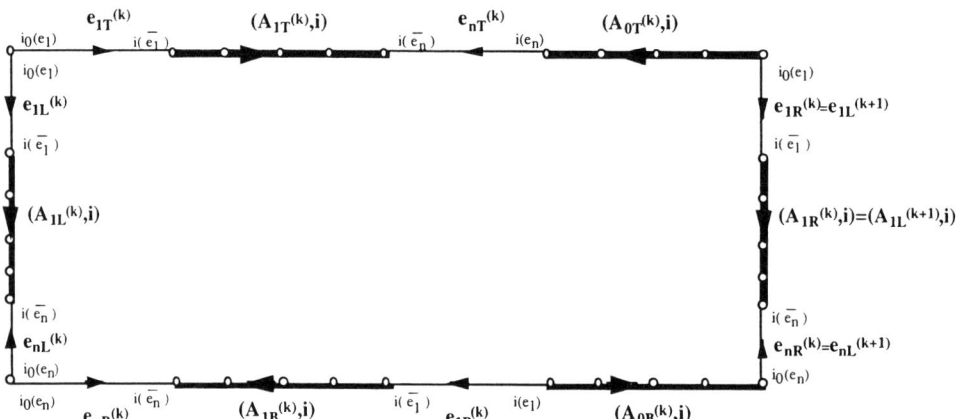

glued together along the connected subgraph $e_{1L}^{(k)} \cup A_{1L}^{(k)} \cup e_{nL}^{(k)}$.

(6) Each $[\frac{d-1}{d}\beta_{2T}]^{(k-1)}$ is replaced by $(d-1)$ copies of $[\frac{1}{d}\beta_{2T}]^{(k-1)}$ with ramification factor 1 (and hence index $i_0([\beta_2])$), attached from $A_{0T}^{(k-1)}$ to the $(d-1)$ copies of $A_{1T}^{(k)}$.

(7) A 'local picture' of the multiple open fanning is given in figure 4.12.7.

(8) We continue to open fan *every occurrence* of $\beta^{(k)}$ in the resulting graph to $(d-1)$ copies of '$\frac{1}{d-1}\beta^{(k)}$'.

(9) The result is an infinite indexed graph (the 'bottom sheet') from which $(d-1)$ sheets ('infinite cusps') branch at the 'left side' of every rectangle.

(10) The indexed graph (B, j) is depicted in figure 4.12.10.

(11) Proposition. *The indexed graph (B, j) is unimodular.*

Proof. The multiple open fanning described above preserves unimodularity of rectangles (this is an iterated application of 4.4.1). We observe also that changing the indices of $e_{1T}^{(k)}$ and $e_{nB}^{(k)}$ from $(d-1)i_0(e_1)$ and $(d-1)i_0(e_n)$ to $i_0(e_1)$ and $i_0(e_n)$ respectively preserves unimodularity of rectangles.

Moreover, the $(d-1)$ rectangles glued together in 4.12.5 are glued along the connected subgraph $e_{1L}^{(k)} \cup A_{1L}^{(k)} \cup e_n^{(k)}$ for each $k = 1, 2, 3, \ldots$ and thus by lemma 4.3.1 such gluings are unimodular. The indexed graph (B, j) is constructed from infinitely many such gluings and from adjoining bridges $[\frac{1}{d}\beta_{2R}]^{(k)}$, $[\frac{1}{d}\beta_{2T}]^{(k)}$, $[\beta_3]^{(k)}$ for $k = 1, 2, 3, \ldots$ By an argument similar to the argument in 4.11 (step 6), adjoining bridges is unimodular and so (B, j) is unimodular. □

(12) Proposition. *The indexed graph (B, j) has bounded denominators.*

Proof. We observe that the projection $p: (B, j) \to (C, j)$ onto the indexed graph (C, j) shown in figure 4.12.12, the 'bottom sheet' of (B, j) is index preserving so we compute in (C, j) for each $s = 1, 2, 3, \ldots$:

$$\frac{\Delta a_{4s+1}}{\Delta a_1} = \frac{d^{s+1}}{d-1}$$

$$\frac{\Delta b_{4s+1}}{\Delta a_1} = \frac{\Delta_{A_0} b_1}{\Delta_{A_0} a_1} \cdot \frac{d^{s+1}}{d-1}$$

$$\frac{\Delta c_{4s+1}}{\Delta a_1} = \frac{\Delta a_{4s+1}}{\Delta a_1} \cdot \frac{i(\overline{e_1})}{i_0(e_1)} = \frac{i(\overline{e_1})}{i_0(e_1)} \cdot \frac{d^{s+1}}{d-1}$$

$$\frac{\Delta d_{4s+1}}{\Delta a_1} = \frac{\Delta b_{4s+1}}{\Delta a_1} \cdot \frac{i(\overline{e_n})}{i_0(e_n)} = \frac{\Delta_{A_0} b_1}{\Delta_{A_0} a_1} \cdot \frac{i(\overline{e_n})}{i_0(e_n)} \cdot \frac{d^{s+1}}{d-1}$$

It follows easily that (C, j) has (BD) and, therefore, that (B, j) has (BD). □

58   LISA CARBONE

(13) Proposition. *The indexed graph (B, j) has finite volume.*

Proof. Choose the vertex
$$a_1 = \partial_0(e^{(0)}_{1_T R}) \in VA^{(0)}_{0R}$$
as the basepoint for the computation of the volume of (B, j). Let V_0 and V_1 be the volumes of the indexed subgraphs (A_0, i) and (A_1, i), computed at the basepoints $\partial_0 e_n \in VA$ and $\partial_1 e_1 \in VA$ respectively; that is:
$$V_0 = Vol_{\partial_0 e_n}(A_0, i),$$
$$V_1 = Vol_{\partial_1 e_1}(A_1, i).$$

We observe that all the vertices of B belong to (a copy of) A_0 or A_1; therefore, our computation of $Vol_{a_1}(B, j)$ will be expressed in terms of V_0 and V_1.

Our strategy is to compute the volume of each 'rectangle' of (B, j) and then obtain an expression for the total volume, by summing the volumes of all such rectangles.

We recall (1.4.12) that changing the basepoint changes the volume computation by a constant, that is : for an indexed graph (A, i) and $a, b \in VA$
$$Vol_a(A, i) = \frac{\Delta a}{\Delta b} Vol_b(A, i).$$

Let $(R^{(-1)}, j) =$

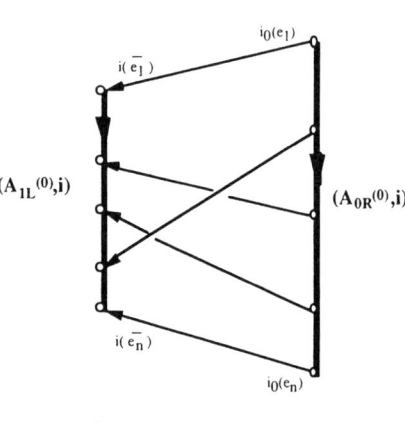

then
$$Vol_{a_1}(R^{(-1)}, j) = Vol_{a_1}(A_0, i) + Vol_{a_1}(A_1, i)$$
$$= \frac{\Delta a_1}{\Delta b_1} V_0 + \frac{\Delta a_1}{\Delta a_0} V_1.$$

We set
$$c_0^{(-1)} = \frac{\Delta a_1}{\Delta b_1},$$
$$c_1^{(-1)} = \frac{\Delta a_1}{\Delta a_0}$$

and thus
$$\boxed{Vol_{a_1}(R^{(-1)}, j) = c_0^{(-1)} V_0 + C_1^{(-1)} V_1.}$$

Let $(R^{(0)}, j) =$
$(\mathbf{R^{(0)}, j}) =$

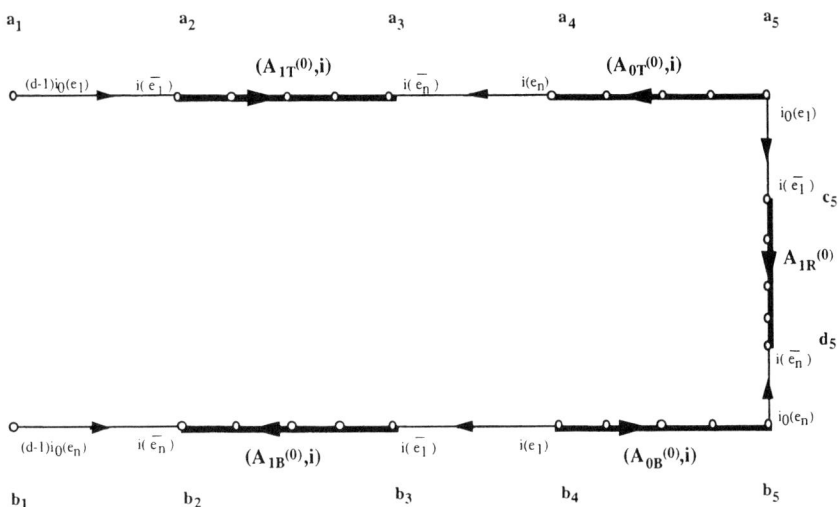

then
$$Vol_{a_1}(R^{(0)}, j) = Vol_{a_1}(A_{1T}^{(0)}, i) + Vol_{a_1}(A_{0T}^{(0)}, i)$$
$$+ Vol_{a_1}(A_{1R}^{(0)}, i) + Vol_{a_1}(A_{0B}^{(0)}, i)$$
$$+ Vol_{a_1}(A_{1B}^{(0)}, i)$$
$$= \left(\frac{\Delta a_1}{\Delta a_4} + \frac{\Delta a_1}{\Delta b_5} + \frac{\Delta a_1}{\Delta b_3}\right) V_1.$$

We set

$$c_0^{(0)} = \frac{\Delta a_1}{\Delta a_4} + \frac{\Delta a_1}{\Delta b_5}$$
$$c_1^{(0)} = \frac{\Delta a_1}{\Delta a_2} + \frac{\Delta a_1}{\Delta c_5} + \frac{\Delta a_1}{\Delta b_3},$$

and we obtain:

$$\boxed{Vol_{a_1}(R^{(0)}, j) = c_0^{(0)} V_0 + c_1^{(0)} V_1.}$$

For $k = 1, 2, \ldots$, let $(R^{(k)}, j) =$

$(\mathbf{R}^{(k)}, \mathbf{j}) =$

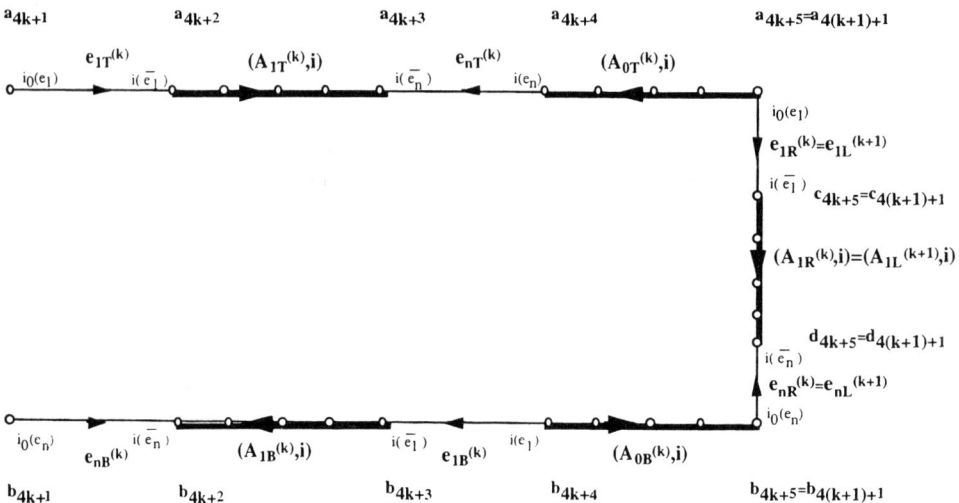

then

$$Vol_{a_{4k+1}}(R^{(k)}, j) = Vol_{a_{4k+1}} 8A_{1T}^{(k)}, i) + Vol_{a_{4k+1}}(A_{0T}^{(k)}, i)$$
$$+ Vol_{a_{4k+1}}(A_{1R}^{(k)}, i) + Vol_{a_{4k+1}}(A_{0B}^{(k)}, i)$$
$$+ Vol_{a_{4k+1}} 8A_{1B}^{(k)}, i)$$

$$= V_0 \left(\frac{\Delta a_{4k+1}}{\Delta a_{4k+4}} + \frac{\Delta a_{4k+1}}{\Delta b_{4k+5}} \right) + V_1 \left(\frac{\Delta a_{4k+1}}{\Delta a_{4k+2}} + \frac{\Delta a_{4k+1}}{\Delta c_{4k+5}} + \frac{\Delta a_{4k+1}}{\Delta b_{4k+3}} \right).$$

For $k = 1, 2, \ldots$ let

$$c_0^{(k)} = \frac{\Delta a_{4k+1}}{\Delta a_{4k+4}} + \frac{\Delta a_{4k+1}}{\Delta b_{4k+5}}$$

$$c_1^{(k)} = \frac{\Delta a_{4k+1}}{\Delta a_{4k+2}} + \frac{\Delta a_{4k+1}}{\Delta c_{4k+5}} + \frac{\Delta a_{4k+1}}{\Delta b_{4k+3}}$$

be the 'change basepoint constants'. Then $c_0^{(k)}$ and $c_1^{(k)}$ are independent of $k = 1, 2, \ldots$ so we denote them c_0 and c_1 respectively.

For $k = 1, 2, \ldots$, we obtain:

$$\boxed{Vol_{a_{4k+1}}(R^{(k)}, j) = c_0^{(k)} V_0 + c_1^{(k)} V_1 = c_0 V_0 + c_1 V_1}$$

In order to change the basepoint for the computation of the volume of each rectangle back to the original basepoint $a_1 \in A_{0R}^{(0)}$, we observe that:

$$\frac{\Delta a_1}{\Delta a_{4k+1}} = \frac{d-1}{d^k}, \qquad k = 1, 2, \ldots$$

The indexed graph (B, j) has one rectangle of type $R^{(-1)}$, one rectangle of type $R^{(0)}$, (d-1) rectangles of type $R^{(1)} \ldots$, and $(d-1)^{(k)}$ rectangles of the type $R^{(k)}$.

Therefore:

$$Vol_{a_1}(B,j) = Vol_{a_1}(R^{(-1)}) + Vol_{a_1}(R^{(0)}) + (d-1)Vol_{a_1}(R^{(1)})$$
$$+ (d-1)^2 Vol_{a_1}(R^{(2)}) + \cdots + (d-1)^k Vol_{a_1}(R^{(k)}) + \ldots$$

$$= (c_0^{(-1)}V_0 + c_1^{(-1)}V_1) + (c_0^{(0)}V_0 + c_1^{(0)}V_1)$$
$$+ (d-1)\frac{\Delta a_5}{\Delta a_1}Vol_{a_1}(R^{(1)}) + (d-1)^2\frac{\Delta a_9}{\Delta a_1}Vol_{a_1}(R^{(2)})$$
$$+ (d-1)^k\frac{\Delta a_{4k+1}}{\Delta a_1}Vol_{a_1}(R^{(k)}) + \ldots$$

$$= (c_0^{(-1)}V_0 + c_1^{(-1)}V_1) + (c_0^{(0)}V_0 + c_1^{(0)}V_1)$$
$$+ (d-1)\frac{(d-1)}{d}(c_0V_0 + c_1V_1) + (d-1)^2\frac{(d-1)}{d^2}(c_0V_0 + c_1V_1)$$
$$+ \cdots + (d-1)^k\frac{(d-1)}{d^k}(c_0V_0 + c_1V_1) + \ldots$$

$$= (c_0^{(-1)} + c_1^{(0)})V_0 + (c_1^{(-1)} + c_1^{(0)})V_1$$
$$+ (d-1)(c_0V_0 + c_1V_1)[\frac{d-1}{d} + (\frac{d-1}{d})^2 + (\frac{d-1}{d})^3 + \ldots]$$
$$< \infty \text{ since } d > 1. \square$$

(14) Proposition. *There is a morphism $p: (B,j) \to (A,i)$ which is a covering of edge-indexed graphs.*

Proof. We exhibit a morphism $p: (B,j) \to (B^-, j^-)$ which is a covering of indexed graphs; then we are done by proposition 4.11.5.

We recall that a general 'rectangle' $R^{(k)}$ in (B,j) is of the form:

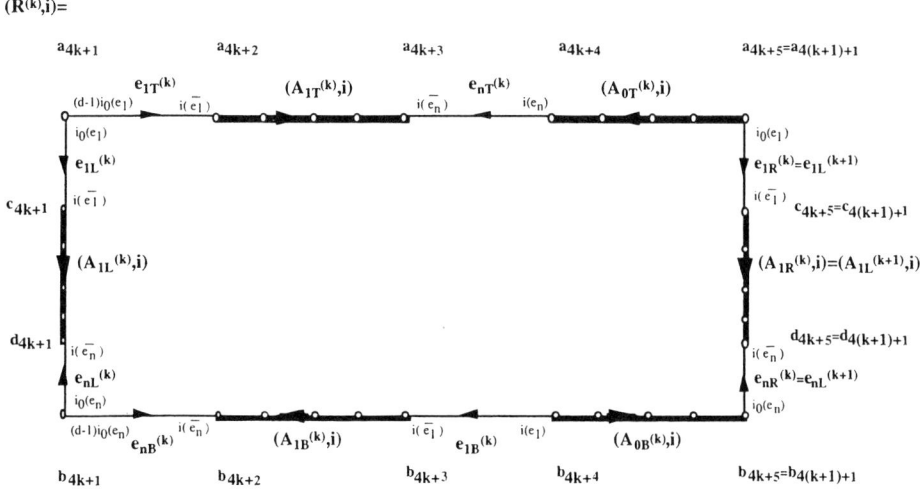

We modify the 'upper label', k of $R^{(k)}$ to indicate the position of $R^{(k)}$ in (B,j): we label a general rectangle in (B,j) as $R^{(kk_1k_2\ldots k_t)}$, $k=1,2,3,\ldots$, $1 \le k_j \le d-1$, $j=1,\ldots,t$, and fig 4.12.14 indicates the position of each rectangle in this labelling scheme.

We extend the 'upper label' $\underline{k} = kk_1k_2\ldots k_t$ of $R^{(kk_1k_2\ldots k_t)}$ to edges and subgraphs within $R^{(kk_1k_2\ldots k_t)}$; that is, we replace the label k on each copy of e_1, e_n, A_0, A_1 in $R^{(kk_1k_2\ldots k_t)}$ by $\underline{k} = kk_1k_2\ldots k_t$ where $k=1,2,3,\ldots$, $1 \le k_j \le d-1$, $j=1,\ldots,t$.

Similarly, we label $[\beta_{2R}]^{(k)}$, $[\beta_{2T}]^{(k)}$, $[\beta_3]^{(k)}$ as $[\beta_{2R}]^{(\underline{k})}$, $[\beta_{2T}]^{(\underline{k})}$, $[\beta_3]^{(\underline{k})}$ respectively.

We define a morphism $p: B \longrightarrow B^-$ as follows:

$$p\mid_{R^{(0)}}: R^{(0)} \xrightarrow{\cong} R^{(0)}$$

For $k = 1, 2, 3, \ldots$, we define:

$$p(e_{1R}^{(k)}) = e_{1R}^{(k)}$$
$$p(e_{nR}^{(k)}) = e_{nR}^{(k)}$$
$$p(e_{1T}^{(k)}) = e_{1T}^{(k)}$$
$$p(e_{nT}^{(k)}) = e_{nT}^{(k)}$$
$$p(e_{1B}^{(k)}) = e_{1B}^{(k)}$$
$$p(e_{nB}^{(k)}) = e_{nB}^{(k)}$$
$$p(e_{1L}^{(k)}) = e_{1L}^{(k)}$$
$$p(e_{nL}^{(k)}) = e_{nL}^{(k)}$$

$$p\,|_{A_{1R}^{(k)}}: A_{1R}^{(k)} \xrightarrow{\cong} A_{1R}^{(k)}$$
$$p\,|_{A_{1T}^{(k)}}: A_{1T}^{(k)} \xrightarrow{\cong} A_{1T}^{(k)}$$
$$p\,|_{A_{0T}^{(k)}}: A_{0T}^{(k)} \xrightarrow{\cong} A_{0T}^{(k)}$$
$$p\,|_{A_{0B}^{(k)}}: A_{0B}^{(k)} \xrightarrow{\cong} A_{0B}^{(k)}$$
$$p\,|_{A_{1L}^{(k)}}: A_{1L}^{(k)} \xrightarrow{\cong} A_{1L}^{(k)}$$
$$p\,|_{A_{1B}^{(k)}}: A_{1B}^{(k)} \xrightarrow{\cong} A_{1B}^{(k)}.$$

For $k = 1, 2, 3, \ldots$ we define:

$$p\,|_{[\frac{1}{d}\beta_{2R}]^{(k)}}: [\frac{1}{d}\beta_{2R}]^{(k)} \xrightarrow{\cong} [\frac{1}{d}\beta_{2R}]^{(k)}$$
$$p\,|_{[\frac{1}{d}\beta_{2T}]^{(k)}}: [\frac{1}{d}\beta_{2T}]^{(k)} \xrightarrow{\cong} [\frac{1}{d}\beta_{2T}]^{(k)}$$
$$p\,|_{[\beta_3]^{(k)}}: [\beta_3]^{(k)} \xrightarrow{\cong} [\beta_3]^{(k)}$$

To verify that the morphism $p\colon (B, j) \to (B^-, j^-)$ is a covering of indexed graphs, we need to compute the sum of the indices in the local fibers above:

(1) $e_{1T}^{(k)}$,

(2) $e_{nB}^{(k)}$;
(3) $[\frac{d-1}{d}\beta_{2T}]^{(k)}$:

(I) The local fiber in $R^{(\underline{k})}$ above $e_{1T}^{(k)}$, $k = 1, 2 \ldots$.

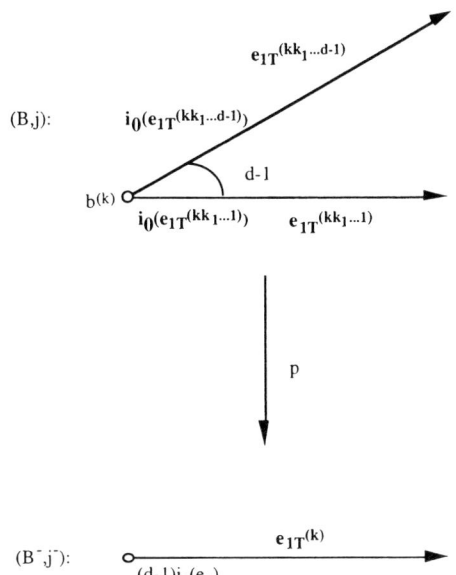

Let $b^{(k)} = p^{-1}(\partial_0(e_{1T}^{(k)}))$. Then:

$$\sum_{f \in p_{b(k)}^{-1}(e_{1T}^{(k)})} j(f) = i_0(e_1) + \cdots + i_0(e_1) \ ((d-1) \text{ times})$$
$$= (d-1)i_0(e_1).$$

(II) The local fiber in $R^{(k)}$ above $e_{nB}^{(k)}$, $k = 1, 2, \ldots$.

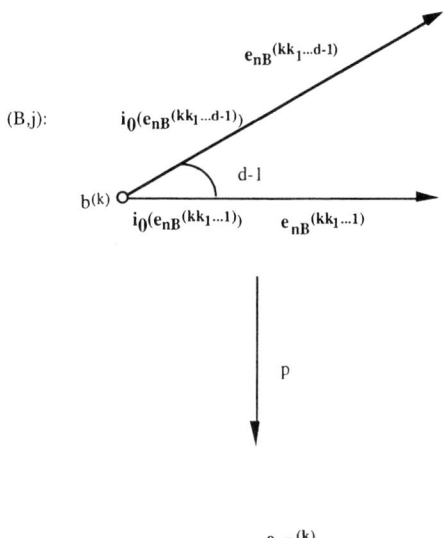

Let $b^{(k)} = p^{-1}(\partial_0(e_{nB}^{(k)}))$. Then:

$$\sum_{f \in p_{r(k)}^{-1}(e_{nB}^{(k)})} j(f) = i_0(e_n) + \cdots + i_0(e_n) \ ((d-1) \text{ times})$$

$$= (d-1)i_0(e_n).$$

(III) The local fiber in $R^{(k)}$ above $[\frac{d-1}{d}\beta_{2T}]^{(k)}$, $k = 1, 2 \ldots$.

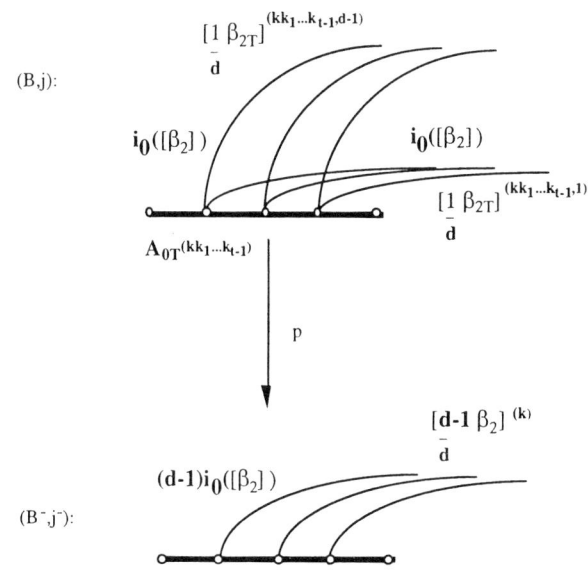

Let $e \in [\frac{d-1}{d}\beta_{2T}]^{(k)}$. Let $b = p^{-1}(\partial_0(e))$. Then:

$$\sum_{f \in p_r^{-1}(e)} j(f) = i_0(e_n) + \cdots + i_0(e_n) \ ((d-1) \text{ times})$$
$$= (d-1)i_0(e_n).$$

For all other edges e with $p^{-1}(\partial_0 e) = b$:

$$\sum_{f \in p_b^{-1}(e)} j(f) = i(e),$$

and thus $p \colon (B, j) \to (B^-, j^-)$ is a covering of indexed graphs. □

4.13 Edge with a common factor implies non-uniform covering

In this section we show that if our edge-indexed quotient graph (A, i) has an edge e whose indices $i(e)$ and $i(\bar{e})$ have a common factor $n > 1$, then (A, i) admits a non-uniform covering.

(1) Theorem (Edge with a common factor implies non-uniform covering). *Let (A, i) be an edge-indexed graph satisfying (F), (U), (MIN), (NDR) and such that there exists $e \in EA$ such that $i(e)$ and $i(\bar{e})$ have a common factor $n > 1$. Then there is a non-uniform covering $p : (B, j) \longrightarrow (A, i)$.*

Proof. If there exists $e \in EA$ such that $i(e)$ and $i(\bar{e})$ have a common factor $n > 1$, then in the barycentric subdivision of (A, i), the edge e *is* an arithmetic bridge for (A, i), and the following (B, j) is a non-uniform covering of (A, i):

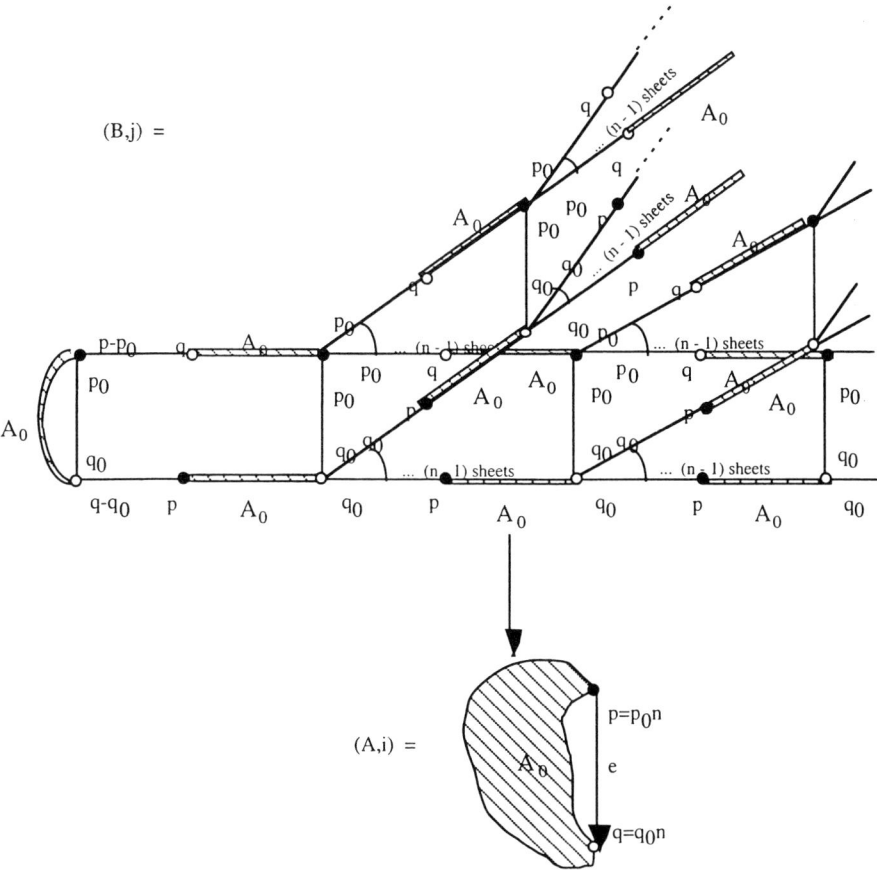

Since the subdivision of the edge e is an arithmetic bridge for (A, i), the properties (INF), (U), (FV) and (BD) for (B, j) are easily verified. □

$(\mathbf{R}^{(\infty)},\mathbf{j}) =$

- fig 4.10.2 -

NON-UNIFORM LATTICES ON UNIFORM TREES

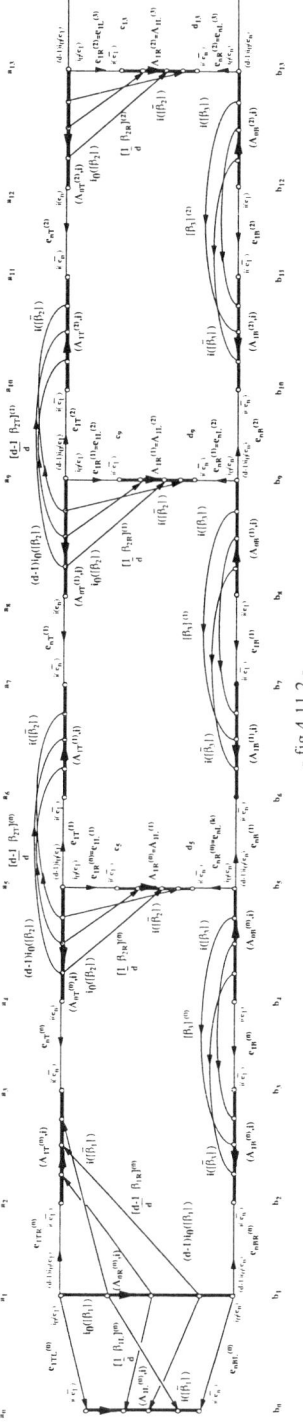

- fig 4.11.2 -

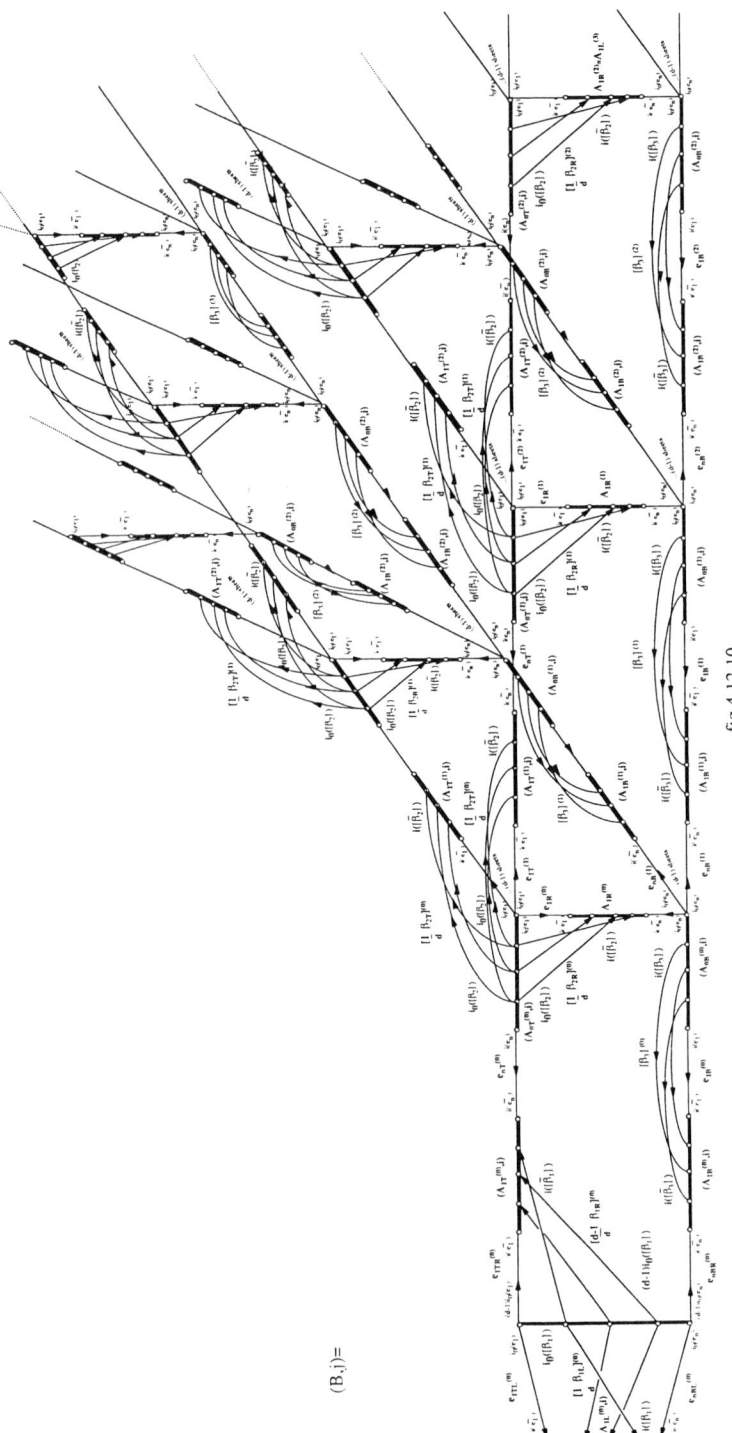

- fig 4.12.10 -

NON-UNIFORM LATTICES ON UNIFORM TREES

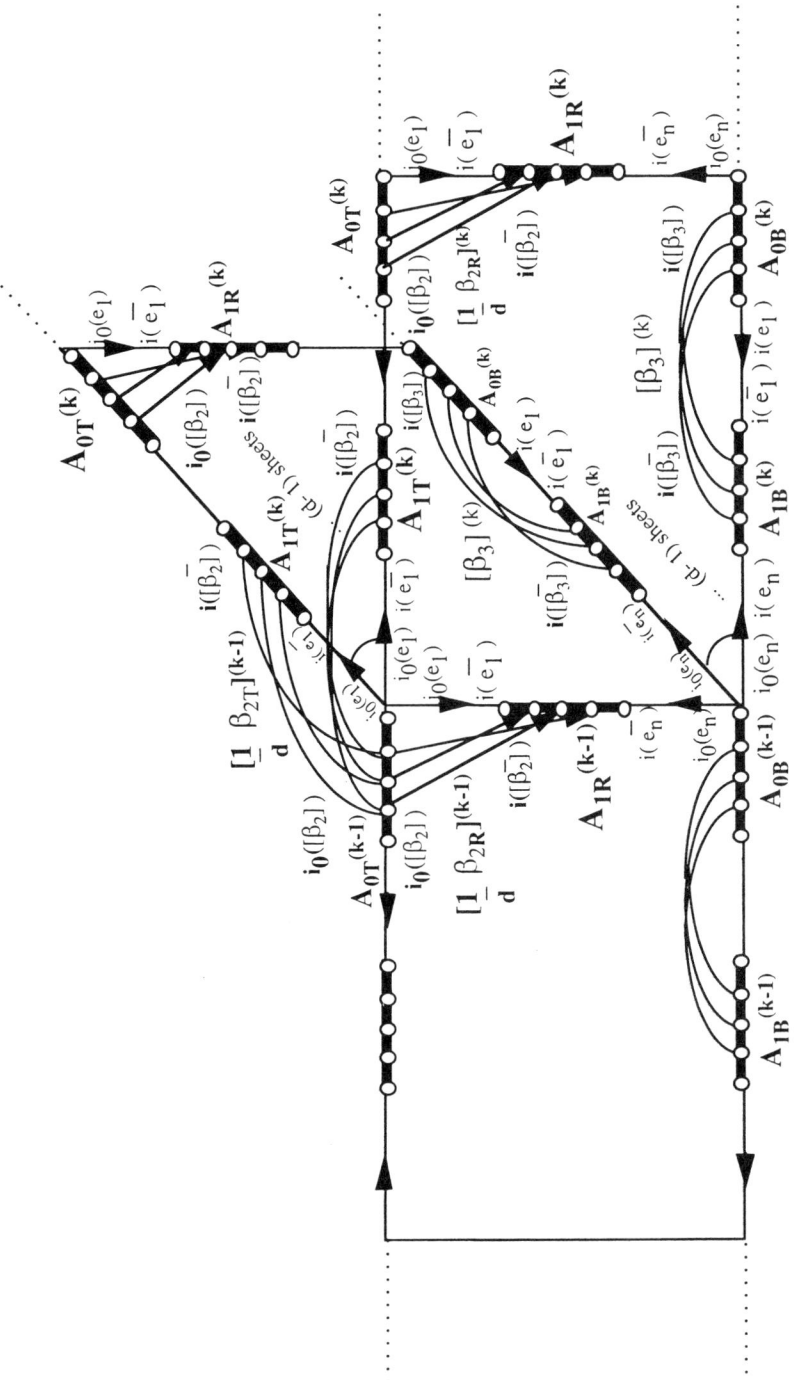

- fig 4.12.7 -

$(C_{i,j}) =$

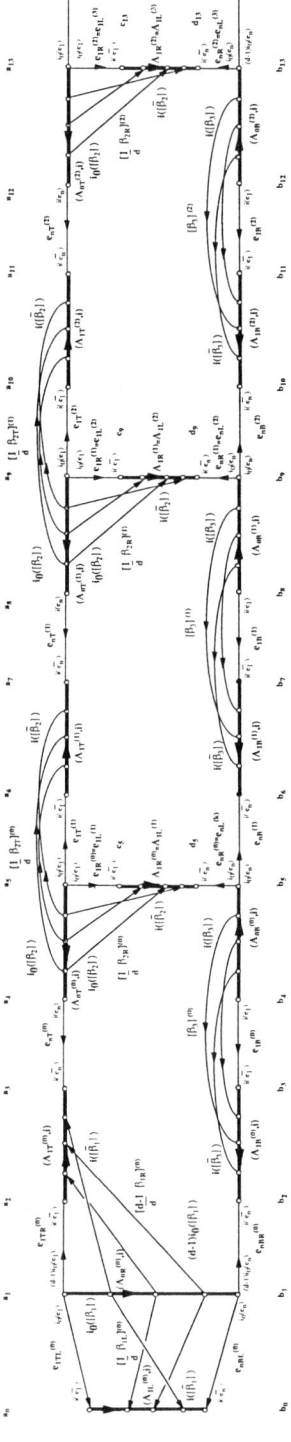

- fig 4.12.12 -

NON-UNIFORM LATTICES ON UNIFORM TREES

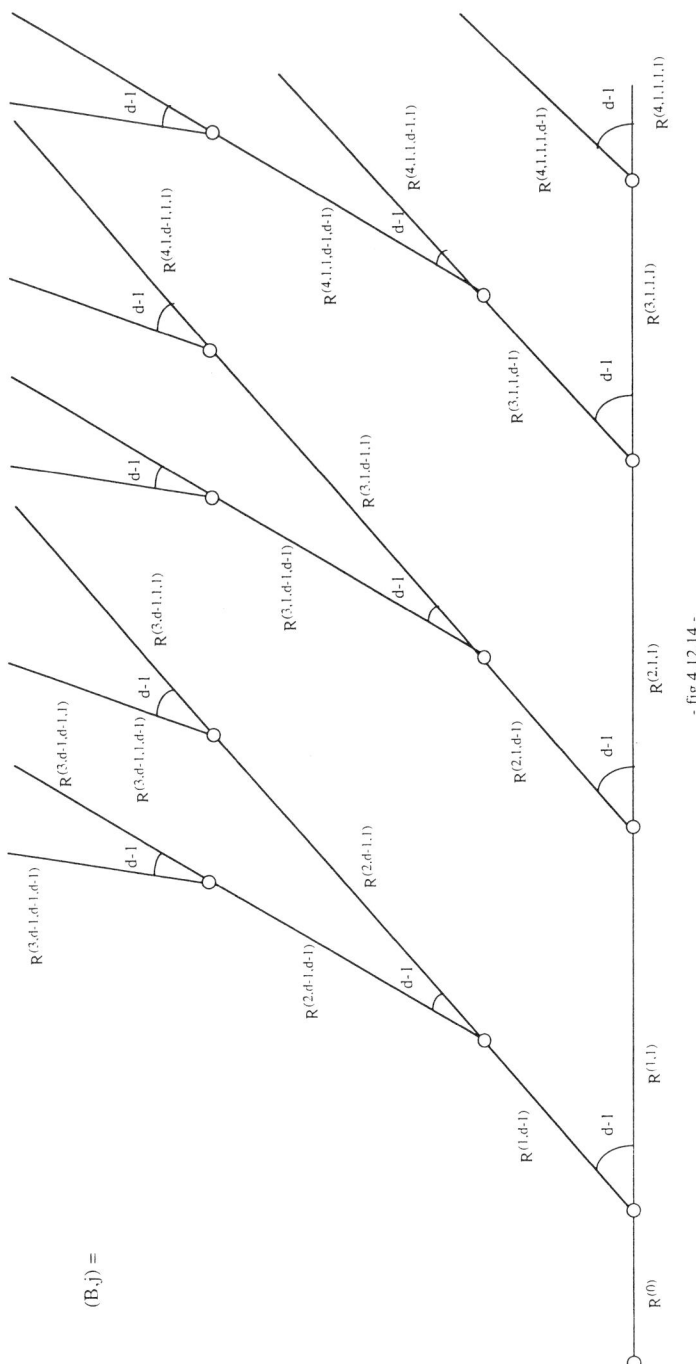

- fig 4.12.14 -

5. Non-uniform coverings of indexed graphs with a ramified separating edge

In this section, we prove conjecture 3.4.15 for edge-indexed graphs that contain a ramified separating edge e:

$$(A, i) = \quad (A_0, i) \xrightarrow{u \quad e \quad v} (A_1, i)$$

that is, $uv \geq 2$. The edge e is an arithmetic bridge for (A, i) (in the sense of definition 4.1.3) consisting of a single edge. The techniques we use to construct non-uniform coverings for (A, i) are different from those in the previous section where we have an arithmetic bridge of two or more edges; we have therefore treated this case separately.

(1) If (A, i) is an indexed graph with a separating edge e, we say that e is a *good separating* for (A, i), if (A, i) is *not* of the form:

$$(A, i) = \quad \circ \xrightarrow{2 \quad e \quad 1} (A_1, i)$$

with $\pi_1(A_1) \neq 1$.

Our objective is to prove:

(2) Theorem (Ramified separating edge implies non-uniform covering). *Let (A, i) be an edge-indexed graph. Suppose that (A, i) satisfies (U), (F), (NDR), (MIN). If (A, i) contains a good ramified separating edge, then (A, i) has a covering $p : (B, j) \longrightarrow (A, i)$ with the properties (U), (INF), (FV), (BD).*

In order to give the proof of this theorem, we treat first the following cases:

(3) For $m \in \mathbb{Z}^+$, let m' denote $m - 1$.

(4) CASE 1. (A, i) *is a (ramified) segment:*

$$(A, i) = \quad \overset{u}{\underset{b_0}{\circ}} \xrightarrow{\hspace{4cm}} \overset{v}{\underset{b_1}{\circ}}$$

(5) We must have $u \geq 2$ and $v \geq 2$ or (MIN) fails, and either $u \geq 3$ or $v \geq 3$ or else (NDR) fails.

(6) We claim that the following is a non-uniform covering of (A, i):

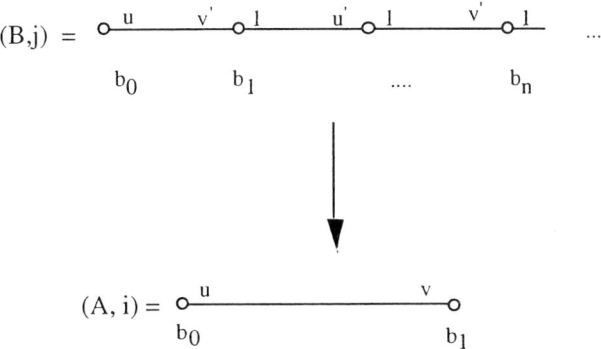

(7) There is an evident covering $p : (B, j) \longrightarrow (A, i)$ ($b_n \in VB$ maps to $b_{(n \bmod 2)}$ in VA). The indexed graph (B, j) clearly satisfies (U) and (INF).

(8) To verify (FV) and (BD), we compute:

$$\frac{\Delta b_{2n}}{\Delta b_0} = \frac{(u'v')^n}{u},$$

$$\frac{\Delta b_{2n+1}}{\Delta b_0} = \frac{\Delta b_{2n}}{\Delta b_0} \cdot v'.$$

(9) It follows that (B, j) satisfies (BD). The volume $Vol_{b_0}(B, j)$ is finite since (MIN) and (NDR) imply that $u'v' > 1$, and so (B, j) satisfies (FV).

(10) CASE 2. *([BL], 8.3(a))*

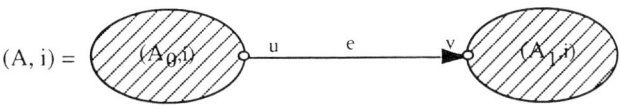

with $u'v' > 1$.

(11) We claim that the following is a non-uniform covering of (A, i):

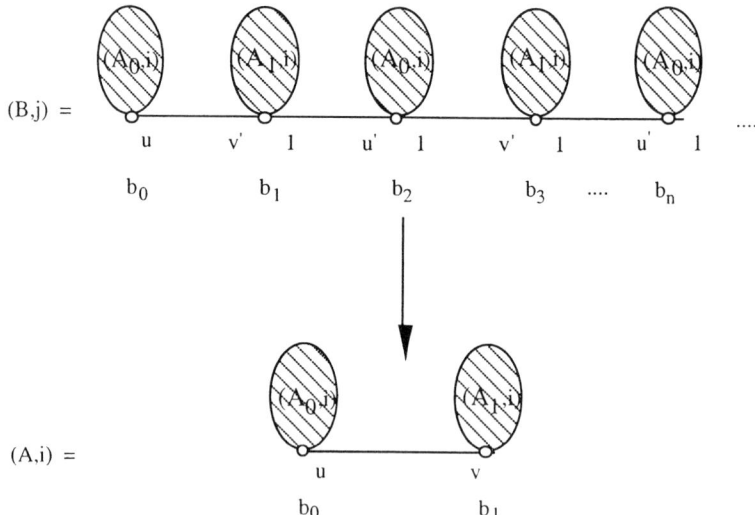

(12) There is an evident covering $p : (B, j) \longrightarrow (A, i)$ ($b_n \in VB$ maps to $b_{(n \bmod 2)}$ in VA). The indexed graph (B, j) clearly satisfies (U) and (INF).

(13) To verify (FV) and (BD), we compute:

$$\frac{\Delta b_{2n}}{\Delta b_0} = \frac{(u'v')^n}{u},$$

$$\frac{\Delta b_{2n+1}}{\Delta b_0} = \frac{\Delta b_{2n}}{\Delta b_0} \cdot v'.$$

(14) It follows that (B, j) satisfies (BD). The volume $Vol_{b_0}(B, j)$ is finite since we assume that $u'v' > 1$, and so (B, j) satisfies (FV).

(15) CASE 3. *([BL], 8.3(b))*

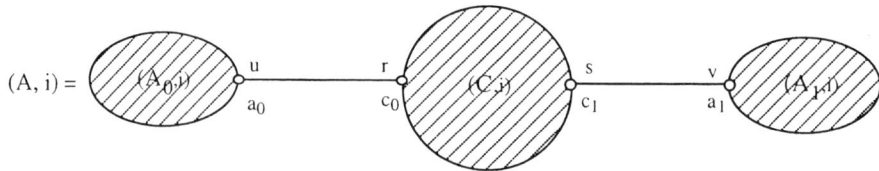

with $u'v' > 1$. We take:

(B,j) =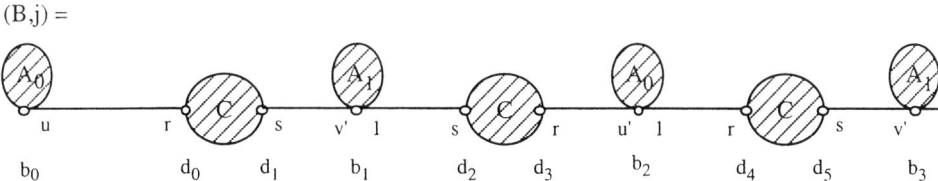

with
$$p(b_j) = a_{j \bmod 2}$$
$$p(d_0) = c_0, \ p(d_1) = c_1, \ p(d_2) = c_1, \ p(d_3) = c_0, \ p(d_4) = c_0, \ p(d_5) = c_1, \ldots$$

(16) It is clear that (B,j) has the properties (U) and (INF). It follows that:

$$\frac{\Delta b_{2n}}{\Delta b_0} = \left(\frac{r}{u}\frac{\Delta c_1}{\Delta c_0}v'\frac{\Delta c_0}{\Delta c_1}u'\right)\left(\frac{r}{1}\frac{\Delta c_1}{\Delta c_0}v'\frac{\Delta c_0}{\Delta c_1}u'\right)\ldots\left(\frac{r}{1}\frac{\Delta c_1}{\Delta c_0}v'\frac{\Delta c_0}{\Delta c_1}u'\right)$$
$$= \frac{(u'v')^n}{u},$$

since $\dfrac{\Delta c_1}{\Delta c_0} = \dfrac{\Delta c_0}{\Delta c_1} = 1$, and we have:

(17) $$\frac{\Delta b_{2n+1}}{\Delta b_0} = \frac{\Delta b_{2n}}{\Delta b_0}\frac{r}{1}\frac{\Delta c_1}{\Delta c_0}\frac{v'}{s}.$$

Thus (B,j) has bounded denominators and has finite volume since $u'v' > 1$.

(18) **CASE 4.** *([BL], 8.4)*

(A, i) =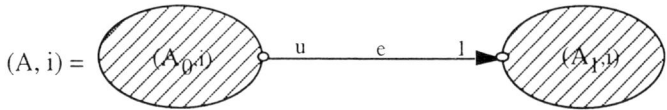

with $u \geq 3$ and $\pi_1(A_1) \neq 1$.

(19) Since $\pi_1(A_1) \neq 1$, A_1 admits a double cover $q : C_1 \longrightarrow A_1$ which we index so that

q is index preserving. Thus we have a covering:

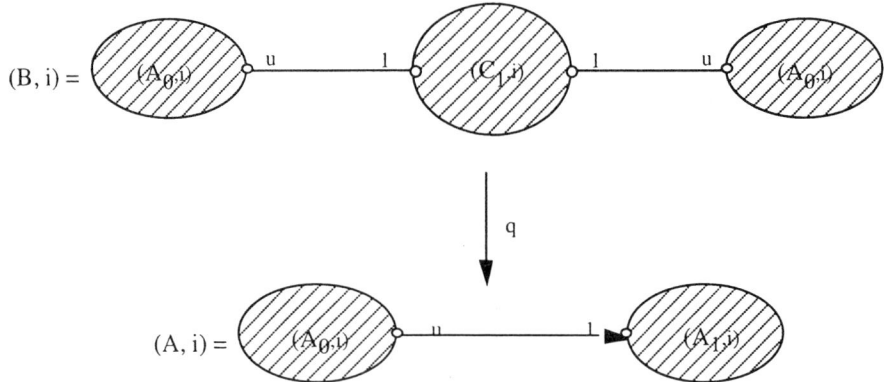

and (B, i) is of type case 3 above.

(20) CASE 5. *([BL], 8.5)*

(a)

$(A, i) =$

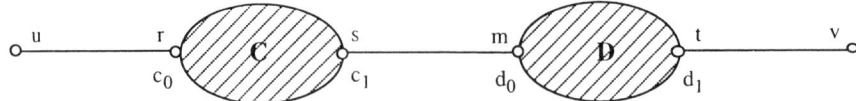

By (MIN), we have $u \geq 2$ and $v \geq 2$. We assume that $m > 1$.

(b)

$(A, i) =$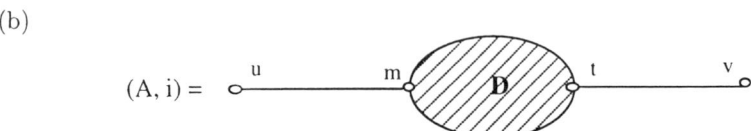

by (MIN), we have $u \geq 2$ and $v \geq 2$, and we assume that $m > 1$.

(c)

$(A, i) =$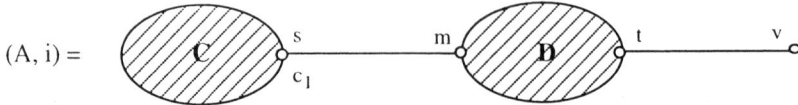

with $v \geq 2$ by (MIN), and $m > 1$.

(21) For (a) we take:

(B,j) =

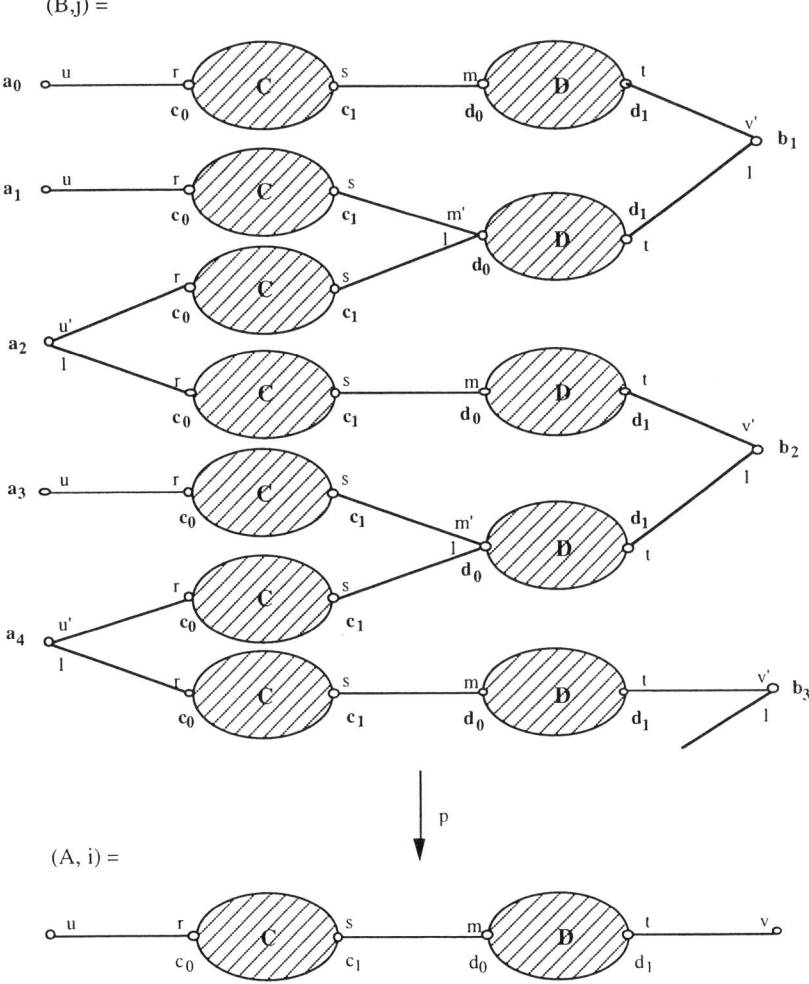

(A, i) =

(22) The indexed graph (B, j) clearly satisfies (INF) and (U), and the covering $p : (B, j) \longrightarrow (A, i)$ is obvious.

(23) Moreover:
$$\frac{\Delta a_{2n}}{\Delta a_0} = \frac{1}{u}(mv'u')^n,$$

$$\frac{\Delta a_{2n-1}}{\Delta a_0} = \frac{mv'}{m'}(mv'u')^{n-1},$$

$$\frac{\Delta b_n}{\Delta a_0} = \frac{\Delta a_{2n}}{\Delta a_0} \frac{r}{1} \frac{\Delta c_1}{\Delta c_0} \frac{m}{s} \frac{\Delta d_1}{\Delta d_0} \frac{v'}{t}.$$

(24) If follows easily that (B,j) has bounded denominators. Since $u'v' \geq 1$ and $m > 1$, it also follows that (B,j) has finite volume.

(25) For (b), observe that there is a covering:

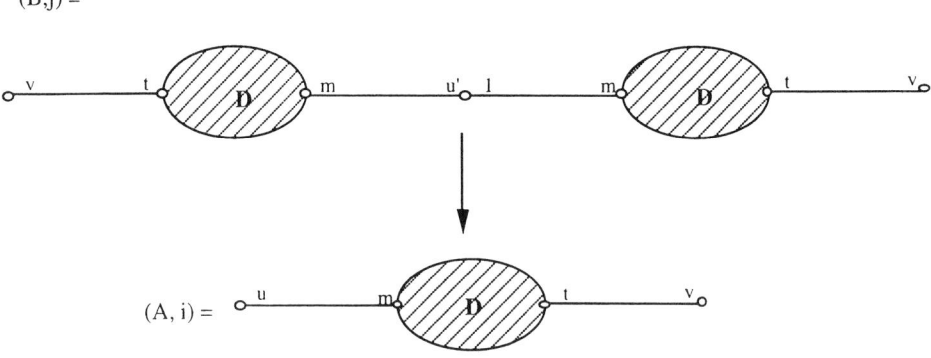

and (B,j) is of type (a) above.

(26) For (c), observe that if $\pi_1(C) \neq 1$, then C admits a double cover $q : C' \longrightarrow C$ which

we index so that q is index preserving. Let $q^{-1}(c_1) = \{b_0, b_1\}$. Then there is a covering:

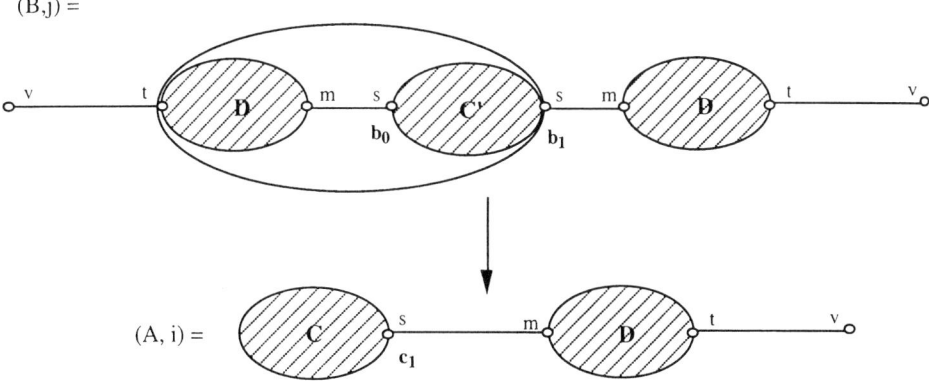

and (B, j) is of type (a) above.

If $\pi_1(C) = 1$, then C is a tree. If $C = \{\circ\}$, then (A, i) is of type (b) above.

If $C \neq \{\circ\}$, then (A, i) is of the form:

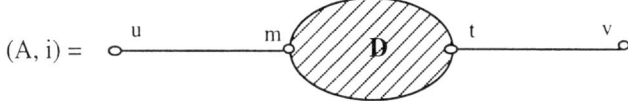

which is of type (b) above.

We are now able to prove:

(27) Theorem (Ramified separating edge implies non-uniform covering). *Let (A, i) be an edge-indexed graph. Suppose that (A, i) satisfies (U), (F), (NDR), (MIN). If (A, i) contains a good ramified separating edge, then (A, i) has a covering $p : (B, j) \longrightarrow (A, i)$ with the properties (U), (INF), (FV), (BD).*

Proof.

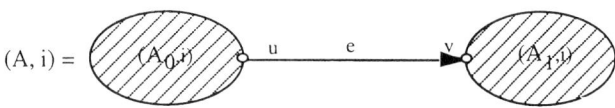

Suppose that (A, i) is a tree. If A_0 and A_1 are single vertices, then we are in case 1 above.

Suppose now that (A,i) is a tree with $A_0 \neq \{\circ\} \neq A_1$. The proof we give in this case is essentially the same as the proof given in [BL], (8.6).

We suppose that for some $e \in EA$, $i(e) = m_0 \geq 3$. Since A is a tree, there is a terminal vertex $a \in VA$, and if $a = \partial_0 f$, we have $i(f) = m_1 \geq 2$ by (MIN).

Choose a path from $\partial_0 e$ to the terminal vertex $a = \partial_0 f$. Then (A,i) is of the form:

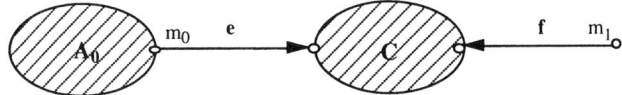

which is covered by case 3 above since $m'_0 m'_1 \geq 2$.

If $i(e) \leq 2$ for each $e \in EA$, then by (NDR), there is an edge e with $i(e) = 2$ and $E_0(\partial_0 e) \neq \{e\}$. We embed e into a path from one terminal vertex of A to another. Then (A,i) is of the form:

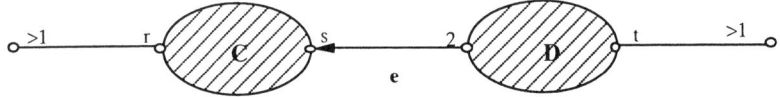

or

which are covered by case 5(a) and case 5(b) respectively.

Suppose now that

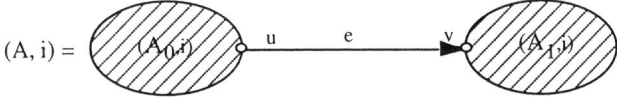

is not a tree. If $u'v' > 1$ then we are in case 2 above.

If $v = 1$, $u \geq 3$, and $\pi_1(A_1) \neq 1$, then we are in case 4 above.

If $v = 1$, $u \geq 3$, and A_1 is a tree, then we can assume $A_1 \neq \{\circ\}$ otherwise we violate (MIN). By (MIN), (A,i) is of the form:

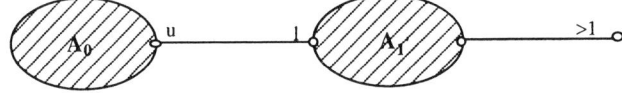

which is of the form case 3 above.

(#) If (A, i) is of the form:

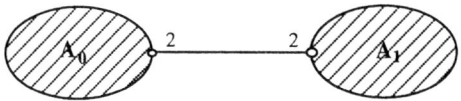

with $\pi_1(A_0) \neq 1$, then A_0 admits a double cover $q: C_0 \longrightarrow A_0$ which we index so that q is index preserving. In this case we have a covering:

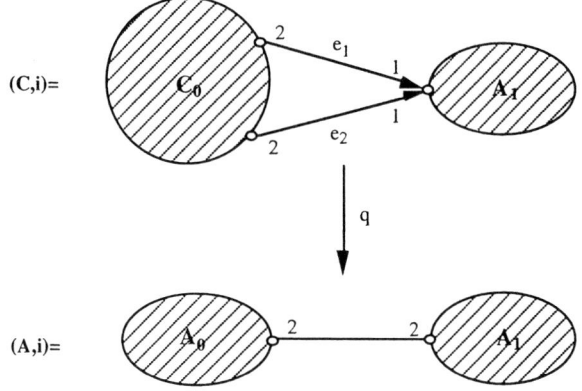

We observe that $\{e_1, e_2\}$ is an arithmetic bridge for (C, i), and thus by theorem 4.5.2 there is a covering $p: (B, j) \longrightarrow (C, i)$ satisfying (U), (INF), (FV), (BD).

If (A, i) is of the form:

and A_0 is a tree, then we can assume that $\pi_1(A_1) \neq 1$, or else A is a tree and we are done. By symmetry, we are in case (#) above.

If

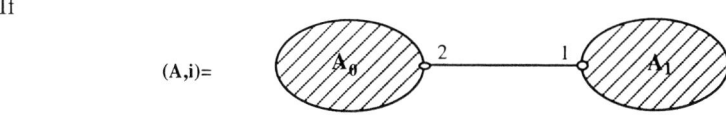

and $\pi_1(A_0) \neq 1 \neq \pi_1(A_1)$, then A_0 and A_1 admit double covers $q_0 : C_0 \longrightarrow A_0$, $q_1 : C_1 \longrightarrow A_1$ which we index so that q_0 and q_1 are index preserving. In this case we have a covering:

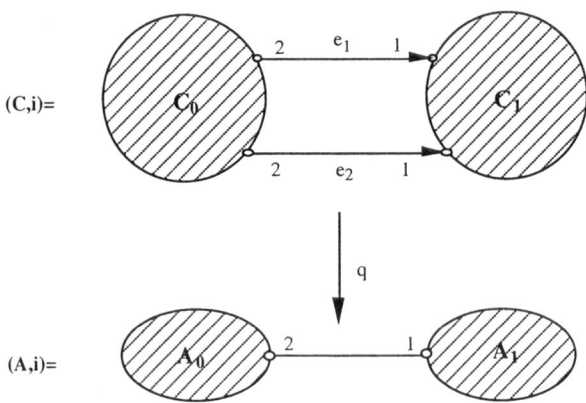

We observe that $\{e_1, e_2\}$ is an arithmetic bridge for (C, i), and thus by theorem 4.5.2 there is a covering $p : (B, j) \longrightarrow (C, i)$ satisfying (U), (INF), (FV), (BD).

If

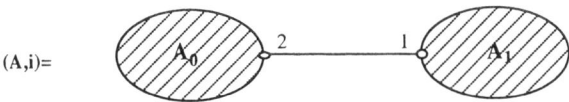

with $\pi_1(A_0) \neq 1$ and A_1 a tree, then A_0 admits an (index preserving) double cover $q : C_0 \longrightarrow A_0$, and (A, i) is covered by:

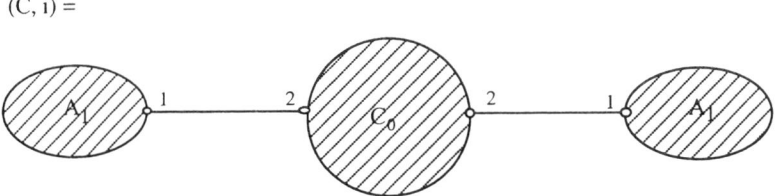

By (MIN), $A_1 \neq \{\circ\}$, and thus (C, i) is of the form:

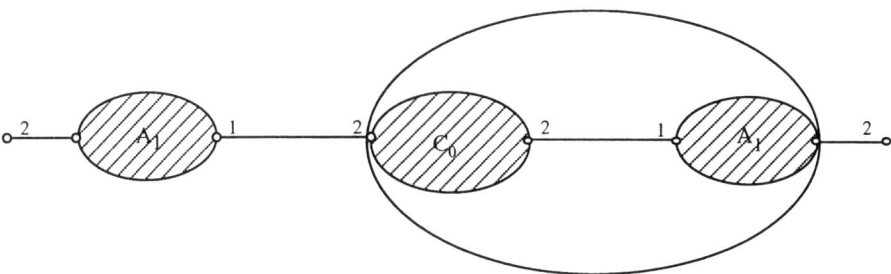

which is of type case 5(a).

If

(A,i)=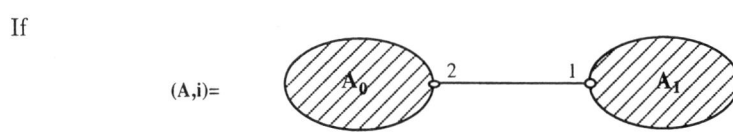

with $\pi_1(A_1) \neq 1$ and A_0 a tree, then we can assume $A_0 \neq \{\circ\}$ otherwise e is not a good separating edge for A. Then A_1 has an (index preserving) double cover $q_1 : C_1 \longrightarrow A_1$, and (A, i) is covered by:

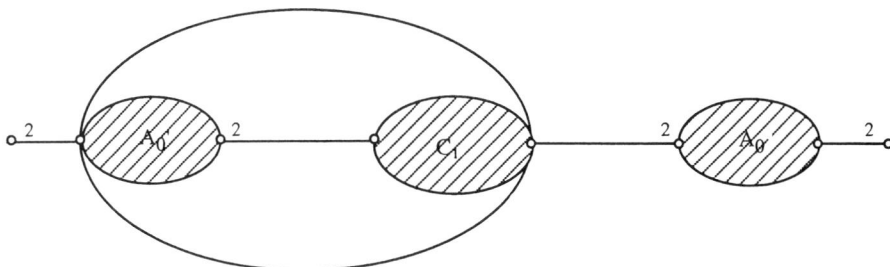

which is of type case 5(a). □

6. Non-uniform coverings of indexed graphs with a ramified loop

(1) The edge-indexed quotient graph of a tree action may contain loops and multiple edges. In the next section, we will describe some machinery for 'eliminating' multiple edges from an indexed graph. In this section, we prove conjecture 3.4.15 for indexed graphs (A, i) with a ramified unimodular loop;

$$L_n = a \bigcirc^n_n$$

where $n \geq 2$. So (A, i) is of the form:

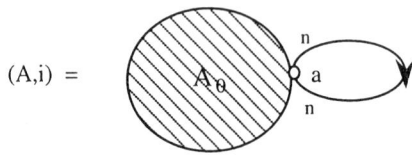

(2) We observe that (L_n, i) automatically satisfies (U), (F), (MIN), and since $n \geq 2$, (L_n, i) satisfies (NDR).

(3) **Lemma (Decorating).** *Let (A, i) be an edge-indexed graph and let $A_0 \subset A$ be a connected subgraph such that A is obtained from A_0 by attaching to each $a \in VA_0$ a copy of a connected edge-indexed graph (A_1, i_1). If (A_0, i_0) satisfies (U) and (FV), and if (A_1, i_1) satisfies (U) and (FV) then (A, i) satisfies (U) and (FV).*

Proof. Since each (A_1, i_1) is attached to (A_0, i_0) at a single vertex, and since (A_0, i_0) and (A_1, i_1) satisfy (U), (A, i) satisfies (U), as a closed path in (A, i) can be written as a product of paths closed in either (A_0, i_0) or (A_1, i_1).

Let $a_0 \in VA_0$ and let $a_1 \in VA_1$ be such that A_1 is attached to A_0 at the vertex a_1. Let $V_0 = Vol_{a_0}(A_0, i_0)$ and let $V_1 = Vol_{a_1}(A_1, i_1)$, then

$$Vol_{a_0}(A, i) = \sum_{a \in VA_0} \frac{V_1}{\left(\frac{\Delta a}{\Delta a_0}\right)}$$
$$= V_1 V_0$$
$$< \infty. \square$$

(4) Corollary:. *Let (A, i) and (A_0, i_0) be as in the lemma, and suppose we have a covering $p_0 : (B_0, j_0) \longrightarrow (A_0, i_0)$ such that (B_0, j_0) satisfies $(INF), (U), (FV)$, and (BD). Let (B, j) be obtained from (B_0, j_0) by attaching to each $b \in VB_0$ a copy of (A_1, i_1), then there is a covering $p : (B, j) \longrightarrow (A, i)$. Furthermore, if (A_1, i_1) is finite, then (B, j) satisfies $(INF), (U), (FV)$, and (BD).*

Proof:. To see that p is a covering of edge-indexed graphs, we extend $p_0 : (B_0, j_0) \longrightarrow (A_0, i_0)$ to $p : (B, j) \longrightarrow (A, i)$ so that p is index-preserving on each attached (A_1, i_1). By the Lemma (Decorating), applied to (B_0, j_0), we have that (B, j) satisfies (U) and (FV). Since (B_0, j_0) satisfies (BD) and (A_1, i_1) is finite, it follows easily that (B, j) satisfies (BD). □

(5) We recall that (A, i) is of the form:

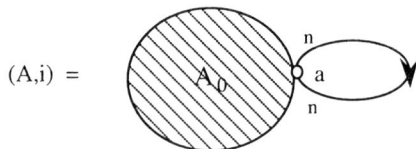

Our strategy is to find a covering $p_n : (C_n, j) \longrightarrow (L_n, i)$ with the properties (INF), (U), (FV), (BD) for the unimodular loop (L_n, i) where $n \geq 2$. By lemma 6.3, we can obtain a covering for (A, i) with the properties (INF), (U), (FV), (BD) by attaching a copy of (A_0, i) to every vertex in $p^{-1}(a)$.

(6) Theorem (Ramified loop implies non-uniform covering). *Let (A, i) be an edge-indexed graph. Suppose that (A, i) satisfies (U), (F), (NDR), (MIN), and that (A, i) contains a ramified loop*

$$L_n = a \overset{n}{\underset{n}{\bigcirc}}$$

where $n \geq 2$. Then (A, i) has a covering $p : (B, j) \longrightarrow (A, i)$ with the properties (U), (INF), (FV), (BD).

Proof. We have $\pi_1(L_n) \neq 1$ and thus L_n admits a double cover $q : D_n \longrightarrow L_n$ which we index so that q is index preserving:

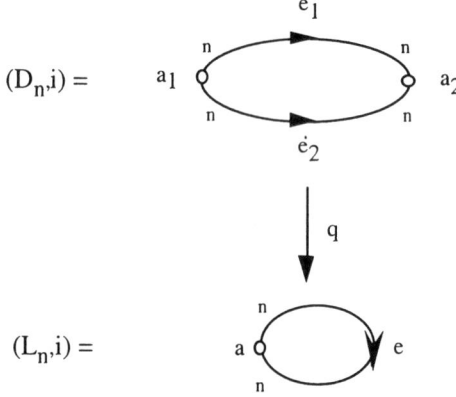

We observe that $\{e_1, e_2\}$ is an arithmetic bridge in (D_n, i) from a_1 to a_2, since $n \geq 2$.

By theorem 4.5.2 (arithmetic bridge implies non-uniform covering), (D_n, i) has a covering $p_n : (C_n, j) \longrightarrow (D_n, i)$ with the properties (INF), (U), (FV), (BD).

The step-by-step construction of non-uniform coverings for indexed graphs with arithmetic bridges (see section 4) applied to (D_n, i) yields the following covering $p_n : (C_n, j) \longrightarrow$

(D_n, i):

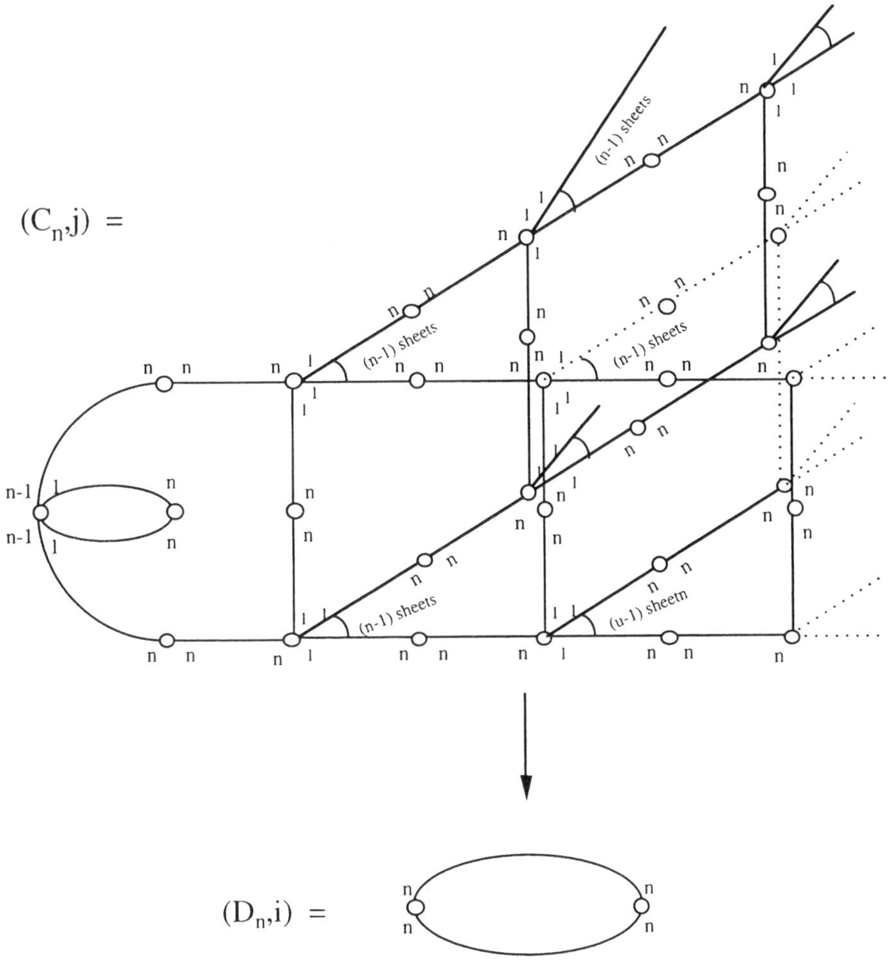

$(C_n, j) = $

$(D_n, i) = $

We recall that (A, i) is of the form:

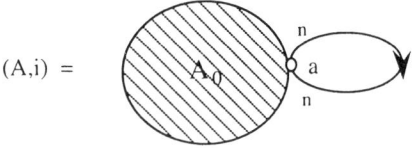

$(A, i) = $

Extending $q : D_n \longrightarrow L_n$ to (A, i) so that q is index preserving on (A_0, i), we obtain a double cover of (A,i):

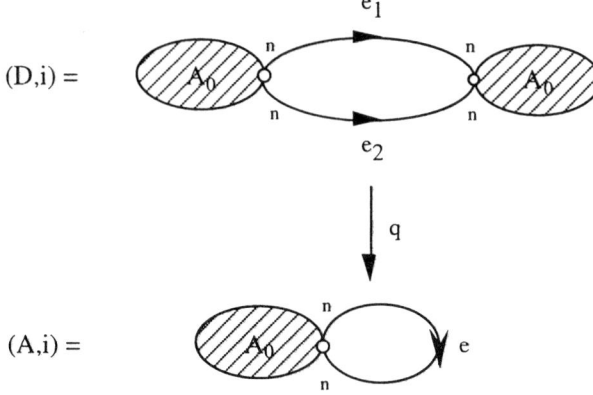

By lemma 6.3 and corollary 6.4, the following is a covering for (A,i) with the properties

(INF), (U), (FV), (BD):

(B,j) =

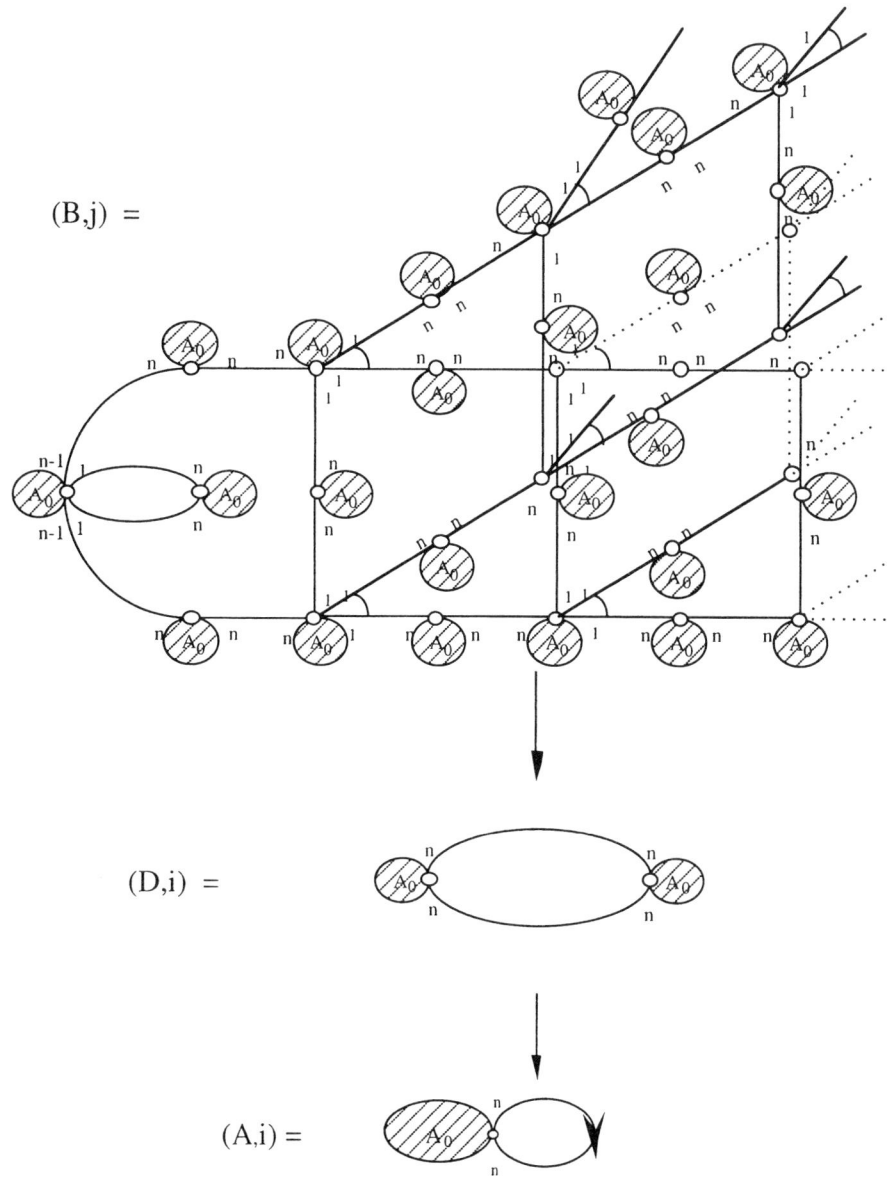

(D,i) =

(A,i) =

□

7. Eliminating multiple edges

The edge-indexed quotient graph (A, i) of a tree action may contain loops and multiple edges. Our techniques for proving 'existence of arithmetic bridges' involve an embedding of (A, i) into a 'complete indexed graph' (see definition 8.2.1). In order to perform such an embedding, we need to ensure that (A, i) has no 'multiple edges'.

Let (A, i) be an indexed graph. For $a, b \in VA$, we denote by $E^A(a, b)$ the set of 'multiple edges' between a and b:

(1) $$E^A(a,b) = \{e \in EA \mid \partial_0 e = a, \partial_1 e = b\}.$$

(2) We define the following conditions on (A, i):

(NL) the graph A has *no loops*.

(RPE) edges in (A, i) are *relatively prime*; that is, for each $e \in EA$ we have

$$gcd(i(e), i(\overline{e})) = 1,$$

and we recall the condition (U); the indexed graph (A, i) is *unimodular*.

(3) We introduce the following constructions in order to treat the situation that the *indexed* quotient (A, i) contains multiple edges:

$$(e^t, i) = \{e_1, \ldots, e_t \mid \partial_0 e_j = a, \partial_1 e_j = b, i(e_j) = u_j, i(\overline{e_j}) = v_j, j = 1, \ldots t\}:$$

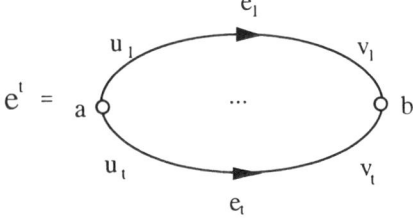

7.1 Simplification of a graph with no loops

(1) Construction (Simplification of a graph with no loops).

Given a graph A with no loops, we define a graph $\underline{A}$ and a projection $q : A \longrightarrow \underline{A}$ as follows:

$$V\underline{A} = VA$$

$$q \mid_{VA} = Id$$

$$E\underline{A} = \{(a,b) \in VA \times VA \mid E^A(a,b) \neq \emptyset\}$$

$$q_{\underline{A}}^{-1}((a,b)) = E^A(a,b), and$$

$$\overline{(a,b)} = (b,a).$$

The graph $\underline{A}$ has no loops or multiple edges, and will be called the *canonical simplification* of A.

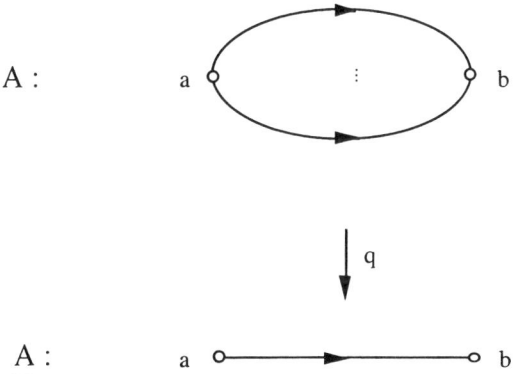

7.2 Graphs with multiplicities

(1) Definition. *A graph with multiplicities,(A, μ), consists of:*

(1) *a graph A;*

(2) *an assignment $\mu : EA \longrightarrow \mathbb{Z}^+$ such that, for every $e \in EA$, $\mu(e) = \mu(\bar{e}) \geq 1$.*

Let (A, μ) be a graph with multiplicities. Suppose that A has no loops, and let $\underline{A}$ be the simplification of A. We define a multiplicity $\underline{\mu}$ on $\underline{A}$ by:

$$\underline{\mu}(a, b) = \sum_{e \in E^A(a,b)} \mu(e), \qquad (2)$$

for each $(a, b) \in V\underline{A}$:

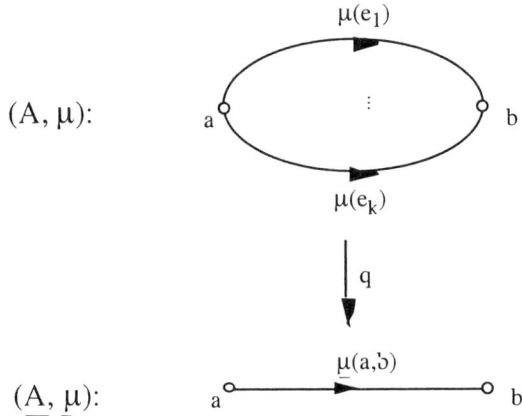

If (A, μ) is a graph with multiplicities, and we have an indexing $i : EA \longrightarrow \mathbb{Z}^+$, then we shall refer to (A, i, μ) as an *indexed graph with multiplicities*.

7.3 Reduced factorization of an indexed graph

(1) Construction (Reduced factorization of an indexed graph).

Let (A, i) be an indexed graph. For each $e \in EA$, write the rational fraction:

$$\frac{i(\bar{e})}{i(e)} = \frac{i_R(\bar{e})}{i_R(e)},$$

in reduced form.

(2) This gives a 'reduced factorization' of i: we have $i(e) = \mu(e) \cdot i_R(e)$, where $gcd(i_R(e), i_R(\bar{e})) = 1$ and $\mu(e) \geq 1$ for each $e \in EA$. Then (A, i_R) is an indexed graph satisfying (RPE) and $i = \mu \cdot i_R$.

(3) Moreover, for each $e \in EA$:

$$\mu(e) = \frac{i(e)}{i_R(e)} = \frac{i(\bar{e})}{i_R(\bar{e})} = \mu(\bar{e}),$$

and thus we have a multiplicity $\mu : EA \longrightarrow \mathbb{Z}^+$ on (A, i_R). The indexed graph (A, i_R, μ) with multiplicities will be called the *reduced factorization* of (A, i).

(4) If (A, i) satisfies (RPE), then $\mu(e) = 1$ for each $e \in EA$, and $i_R(e) = i(e)$ for each $e \in EA$.

7.4 Canonical simplification of a unimodular indexed graph with no loops

(1) Construction (Canonical simplification of a unimodular indexed graph with no loops.).

Let (A, i) be a unimodular indexed graph with no loops, and with reduced factorization (A, i_R, μ). Let $a, b \in VA$. Then for $e \in E^A(a, b)$

$$\frac{i(\bar{e})}{i(e)} \ (= \frac{i_R(\bar{e})}{i_R(e)})$$

depends only on (a, b), not on e, by unimodularity of (A, i). Thus $i_R(e)$ depends only on (a, b).

(2) We define $\underline{i}(a, b) = i_R(e)$, and this defines an indexing $\underline{i}$ on the simplification $\underline{A}$ of A.

(3) We recall that the multiplicity μ on A induces a multiplicity $\underline{\mu}$ on $\underline{A}$:

$$\underline{\mu}(a, b) = \sum_{e \in E^A(a,b)} \mu(e),$$

for each $(a,b) \in V\underline{A}$:

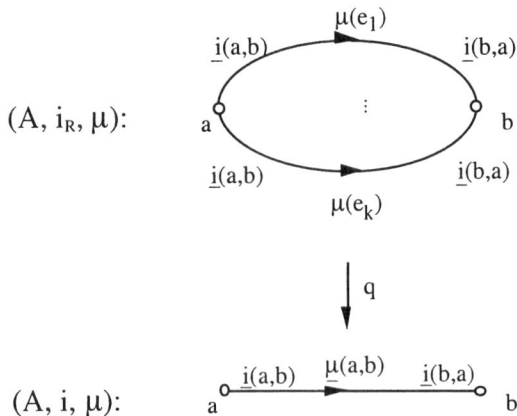

(4) Then $(\underline{A},\underline{i},\underline{\mu})$ is an indexed graph with multiplicities and no multiple edges satisfying (U) and (NL). Let $a, b \in VA$ and $e \in E^A(a,b)$. We have

$$gcd(\underline{i}(a,b), \underline{i}(b,a)) = gcd(i_R(e), i_R(\overline{e})) = 1,$$

so $(\underline{A},\underline{i},\underline{\mu})$ satisfies (RPE). The indexed graph $(\underline{A},\underline{i},\underline{\mu})$ with multiplicities will be called the *canonical simplification* of (A,i) (or of (A, i_R, μ)).

(5) Example.

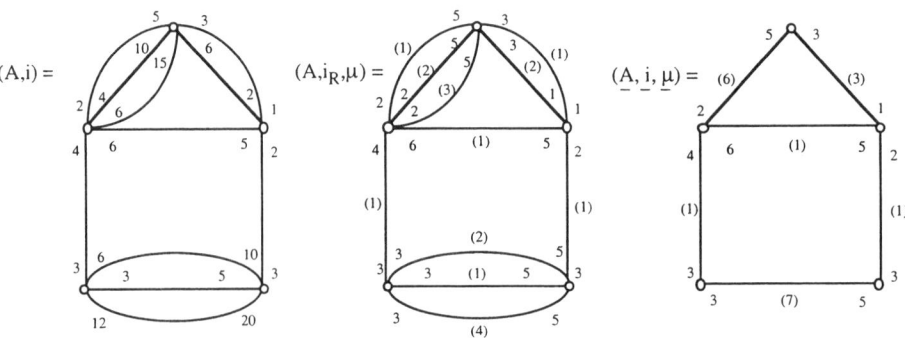

(6) Lemma. *The projection $q : (A, i) \longrightarrow (\underline{A}, \underline{\mu} \cdot \underline{i})$ is a covering of indexed graphs.*

Proof. For $(a, b) \in E\underline{A}$, observe that $q^{-1}(a,b) = E^A(a,b)$:

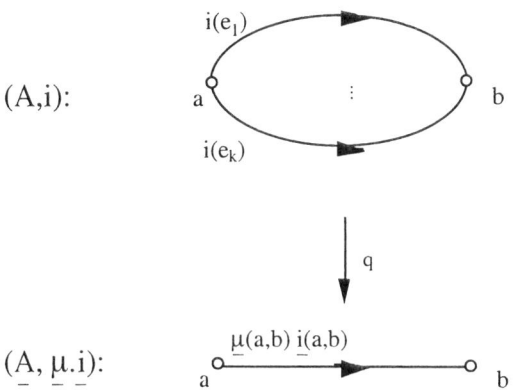

We have:

$$\begin{aligned}
(\underline{\mu} \cdot \underline{i})(a,b) &= \underline{\mu}(a,b) \cdot \underline{i}(a,b) \\
&= \sum_{e \in E^A(a,b)} \mu(e)\underline{i}(a,b) \\
&= \sum_{e \in E^A(a,b)} \mu(e) i_R(e) \\
&= \sum_{e \in E^A(a,b)} i(e). \square
\end{aligned}$$

If (A, i) satisfies (F) and (U), then $(\underline{A}, \underline{i}, \underline{\mu})$ automatically satisfies (F) and (U).

If (A, i) satisfies (RPE), then $i_R(e) = i(e)$ for each $e \in EA$, and the canonical simplification $(\underline{A}, \underline{i}, \underline{\mu})$ is obtained from (A, i) by replacing each multiple edge $E^A(a, b)$ by a single edge (a, b) with multiplicity $\underline{\mu}(a, b)$, where

$$\underline{\mu}(a,b) = |E^A(a,b)|.$$

If (A, i) satisfies (F), (U), (NDR), (MIN) and (RPE) then $(\underline{A}, \underline{i}, \underline{\mu})$ may possibly violate (MIN).

Say that (A,i) is *weakly ramified* if $i(e) \leq 2$ for each $e \in EA$, and *strongly ramified* otherwise.

If (A,i) is weakly ramified, and (A,i) satisfies (F), (U), (NDR), (MIN) and (RPE) then $(\underline{A},\underline{i},\underline{\mu})$ may possibly violate (NDR), but if (A,i) is strongly ramified, then $(\underline{A},\underline{i},\underline{\mu})$ automatically satisfies (NDR).

We refer the reader to §8 for a detailed discussion of properties of $(\underline{A},\underline{i},\underline{\mu})$.

8. Existence of arithmetic bridges

Theorems 4.5.2 and 5.27 allow us to find non-uniform coverings for indexed graphs that contain an arithmetic bridge with $n \geq 2$ edges, or a 'good' separating edge respectively. In this section we complete the proof of conjecture 3.4.13 for the existence of non-uniform lattices on uniform trees; that is, in view of theorems 4.5.2 and 5.27, we prove a theorem establishing the existence of arithmetic bridges in indexed graphs satisfying our hypotheses.

The author would like to thank G. Rosenberg for his careful reading of this section, and for some important corrections. Rosenberg has also pointed out (see [R]) that taking a barycentric subdivision (A_S, i_S) of our edge-indexed quotient graph (A, i) allows us to assume that (A_S, i_S) has no multiple edges, has relatively prime edges, and no loops. This will give an alternate way to prove the results in this section without having to pass to the canonical simplification $(\underline{A}, \underline{i}, \underline{\mu})$.

8.1 Unramified loops

(1) We recall from 1.3.2 that an indexed graph (A, i) is *unramified* if:

$$i(e) = 1 \text{ for every } e \in EA.$$

Let (UL) denote the condition on (A, i):

(UL) All loops of A are unramified; that is, if $e \in EA$ is a loop, then

$$i(e) = i(\overline{e}) = 1.$$

(2) Following [BL], [BT], we say that (A, i) is *discretely ramified* if for $e \in EA$

(DR) $i(e) > 1 \implies i(e) = 2$, e is separating, and $(A_0(e), i)$ is an unramified tree:

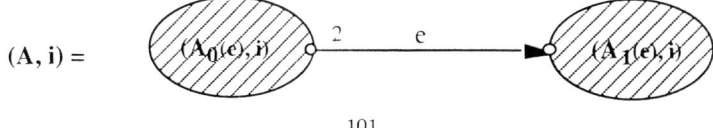

In the presence of condition (MIN), this condition simplifies to:

$(DR)_{min}$

$i(e) > 1 \implies i(e) = 2$, e is separating, and $\partial_0 e$ is a geometrically terminal vertex.

The negations of these conditions give condition *non-discretely ramified* for (A, i):

(NDR) There exists $e \in EA$ such that $i(e) \geq 3$, or $i(e) = 2$ and e is not separating, or $i(e) = 2$, and $(A_0(e), i)$ is either a ramified tree, or an unramified graph.

and in the presence of (MIN) this simplifies to:

$(NDR)_{min}$ There exists $e \in EA$ such that $i(e) \geq 3$, or $i(e) = 2$ and $\partial_0 e$ is not a geometrically terminal vertex.

By Theorems 4.5.2, 4.13.1, 5.27, 6.6, we have

(3) Theorem (Existence of non-uniform coverings).

Let (A, i) be an edge-indexed graph. Suppose that (A, i) satisfies (U), (F), (NDR), (MIN). If (A, i) contains an arithmetic bridge with $n \geq 2$ edges, an edge whose indices have a common factor $n > 1$, a 'good' separating edge or a ramified loop, then (A, i) has a non-uniform covering.

Our aim is to prove the following theorem, and then our proof of conjecture 3.4.13 will be complete.

(4) Theorem (Existence of arithmetic bridges). *Let (A, i) be an edge-indexed graph. Suppose that (A, i) satisfies (U), (F), (NDR), (MIN), (RPE) and (UL). Then (A, i) contains either an arithmetic bridge with $n \geq 2$ edges, or a 'good' separating edge.*

Proof. Let A' denote the graph A with its (unramified) loops removed. We represent A schematically as a core graph A_0, which has no loops or geometrically terminal vertices, to which are attached (sub)-trees (possibly trivial) which connect the core A_0 to the loops of A. The core A_0 may consist of a single vertex (in which case A' is a tree) and in general is obtained from A' by successively 'pruning' all geometrically terminal edges.

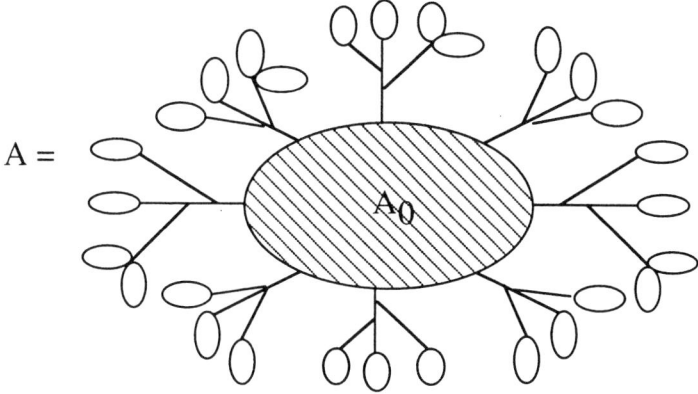

We observe that (A', i) may violate (MIN). Suppose that $A' = A-$loops satisfies (DR). Then

(a) (A', i) is of the form

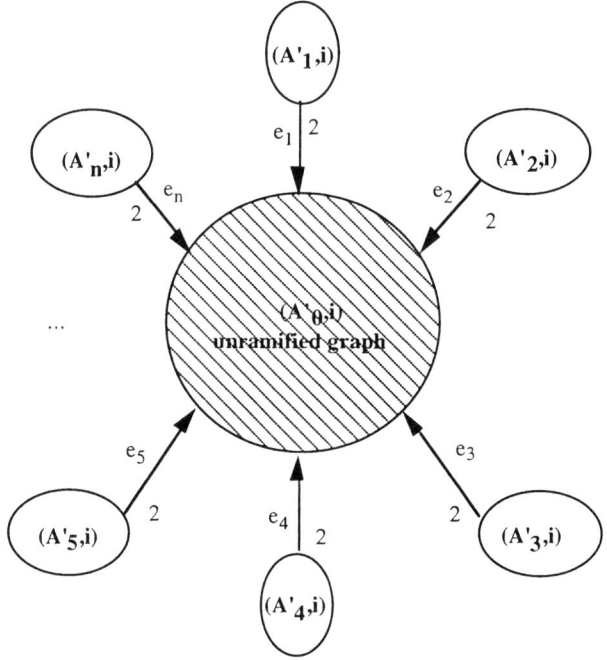

where A_0' is an unramified graph, and each of the (A_j', i) $j = 1, \ldots, n$ are unramified trees.

or

(b) (A', i) is of the form

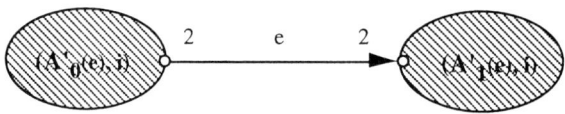

where $(A_j'(e), i)$, $j = 0, 1$ are both unramified trees.

Suppose we are in case (a). Since $A = A' \cup loops$ satisfies (MIN), if $A_j' \neq \{\circ\}$ for some $j = 1, \ldots, n$, then A_j' must have (a bouquet of) unramified loops attached to every terminal vertex of A_j'. It follows that e_j is a good separating edge for (A, i), and we are done. Suppose, conversely, that $A_j' = \{\circ\}$ for every $j = 1, \ldots, n$. Since $A = A' \cup loops$ satisfies (NDR), and since A_0' and all loops are unramified, it must be the case that for some $j = 1, \ldots, n$, $\partial_0 e_j$ is not a terminal vertex of (A, i), that is, e_j has (a bouquet of) unramified loops attached at $\partial_0 e_j$. It follows that e_j is a good separating edge for (A, i), and we are done in case (a).

Suppose we are in case (b). The same argument as above in case (a) (where (A_j', i) $j = 1, \ldots, n$ is replaced by $(A_j'(e), i)$, $j = 0, 1$) shows that e must be a good separating edge for (A', i).

We can therefore assume that (A', i) satisfies (NDR). Thus (A', i) contains a ramified edge.

We consider two possibilities, first that the core (A_0, i) of (A', i) is unramified, and that the ramified edge e belongs to one of the trees connecting the core (A_0, i) to the loops of (A, i). In this case, we shall argue that the ramified edge e is a good separating edge for (A, i).

The second possibility is that the ramified edge e belongs to the core (A_0, i). In this case, we shall argue that e is either a good separating edge for (A, i), or is contained in an arithmetic bridge for (A, i).

In the first case, we assume that the core (A_0, i) of (A', i) is unramified, but (A, i) contains a ramified edge e:

(1) $(A, i) =$ [diagram: (A'_0, i) —>1— e —→ (A_1, i)]

(2) $(A, i) =$ [diagram: (A'_0, i) — e —>1→ (A_1, i)]

where A'_0 contains the core A_0 of A' and A_1 may consist of a single vertex or contain a bouquet of unramified loops.

In case (1) above, if $i(e) > 2$, or if $A'_0 \neq \{\circ\}$, then e is a good separating edge for (A, i).

Suppose that in case (1) $A'_0 = \{\circ\}$ and $i(e) = 2$ (this includes the case that the core A_0 of A' is a single vertex in which case A' is a tree):

$(A, i) =$ [diagram: $\circ$ —2— e —m→ (A_1, i)]

If $m \geq 2$ then e is a good separating edge for (A, i). (If $m = 2$ then $A_1 \neq \{\circ\}$ or (A, i) violates (NDR)).

If $m = 1$, then e is not a good separating edge, however, (A_1, i) cannot be unramified otherwise (A, i) violates (NDR). Moreover $A_1 \neq \{\circ\}$ or (A, i) violates (MIN). Since all loops contained in A_1 are unramified, A_1 must contain a ramified edge f, which moreover, will be separating:

(a) $(A, i) =$ [diagram: $\circ$ —2— e —1→ (A'_1, i) —>1— f —→ (A''_1, i)]

(b) $(A, i) =$ [diagram: $\circ$ —2— e —1→ (A'_1, i) —s— f —>1→ (A''_1, i)]

In (a), f is a good separating edge for (A, i). Suppose we are in case (b). If $i(\overline{f}) > 2$, or if $A_1'' \neq \{\circ\}$, then $\overline{f}$ is a good separating edge for (A, i). Suppose $A_1'' = \{\circ\}$ and $i(\overline{f}) = 2$. If $s \geq 2$ then f is a good separating edge for (A, i). Suppose $s = 1$:

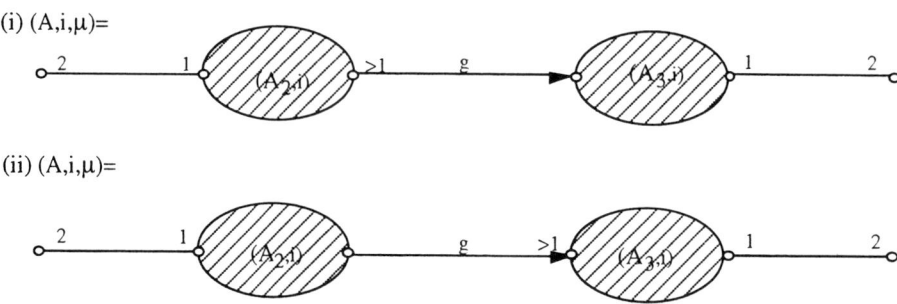

$(A, i, \mu) =$

Then A_1' must contain a ramified edge g which moreover, will be separating:

(i) $(A, i, \mu) =$

(ii) $(A, i, \mu) =$

In case (i), g is a good separating edge for (A, i), as is $\overline{g}$ in (ii).

This completes (1). Suppose now we are in case (2) above:

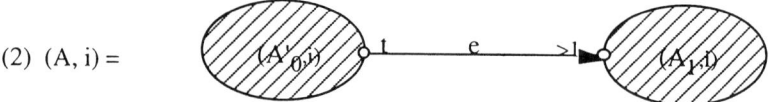

(2) $(A, i) =$

where A_0' contains the core A_0 of A' and A_1 may consist of a single vertex or contain a bouquet of unramified loops.

If $A_1 \neq \{\circ\}$ or if $i(\overline{e}) > 2$, then $\overline{e}$ is a good separating edge for (A, i). If $A_1 = \{\circ\}$ and $t \geq 2$, then then e is a good separating edge for (A, i). If $t = 1$ and $i(\overline{e}) = 2$, then by symmetry, we can repeat the above argument and find a good separating edge for (A, i). This completes the case that the core A_0 of A' is unramified and there is a ramified edge in one of the (sub)-trees connecting A_0 to the (unramified) loops of A.

We now assume that the core A_0 of A' is ramified and that $\pi_1(A_0) \neq 1$ (that is, $A_0 \neq \{\circ\}$.) Then (A_0, i) automatically satisfies (F), (U), is geometrically minimal (has no terminal vertices), so it satisfies (MIN), and since (A_0, i) is ramified and has no terminal vertices, it satisfies (NDR). We observe also that (A_0, i) has no loops. Our remaining objective is to prove (see Corollary 8.4.4):

Theorem (Existence of arithmetic bridges in the core). *Let (A, i) be an indexed graph satisfying (F), (U), (NDR), (MIN), (RPE), and suppose that A has no loops. Then (A, i) contains an arithmetic bridge with $n \geq 2$ edges, or a 'good' separating edge.*

With the proof of this theorem, we are done, since an arithmetic bridge or separating edge for the core (A_0, i) is an arithmetic bridge or separating edge for (A, i).

Our techniques for proving the theorem involve an embedding of (A, i) into a 'complete indexed graph' (see definition 8.2.1). In order to perform such an embedding, we need to ensure that (A, i) has no 'multiple edges'. We can achieve this by replacing (A, i) by its *canonical simplification* $(\underline{A}, \underline{i}, \underline{\mu})$ defined in 7.4.4.

Let (A, i) be an indexed graph satisfying (F), (U), (NDR), (MIN), (RPE), and suppose that A has no loops.

We recall that the graph $\underline{A}$ satisfies (NL), that is, $\underline{A}$ has *no loops*, and $(\underline{A}, \underline{i}, \underline{\mu})$ satisfies (RPE). If (A, i) satisfies (F) and (U), then $(\underline{A}, \underline{i}, \underline{\mu})$ automatically satisfies (F) and (U).

If (A, i) satisfies (RPE), then $i_R(e) = i(e)$ for each $e \in EA$, and the canonical simplification $(\underline{A}, \underline{i}, \underline{\mu})$ is obtained from (A, i) by replacing each multiple edge $E^A(a, b)$ by a single edge (a, b) with multiplicity $\underline{\mu}(a, b)$, where

$$\underline{\mu}(a, b) = |E^A(a, b)|.$$

If (A, i) satisfies (F), (U), (NDR), (MIN) and (RPE) then $(\underline{A}, \underline{i}, \underline{\mu})$ may possibly violate (MIN). However, this will be of no consequence in the applications (see Corollary 8.4.4).

Say that (A, i) is *weakly ramified* if $i(e) \leq 2$ for each $e \in EA$, and *strongly ramified* otherwise. If (A, i) is weakly ramified, and (A, i) satisfies (F), (U), (NDR), (MIN) and (RPE) then $(\underline{A}, \underline{i}, \underline{\mu})$ may possibly violate $(NDR)_{min}$ (once again, this will be of no consequence in the applications (see Corollary 8.4.4)), but if (A, i) is strongly ramified, then $(\underline{A}, \underline{i}, \underline{\mu})$ automatically satisfies $(NDR)_{min}$.

8.2 Completion

In this section, we complete the first step of the proof of theorem 8.1.4 by embedding the canonical simplification $(\underline{A}, \underline{i}, \underline{\mu})$ into a 'complete indexed graph' with multiplicities.

Let (A, i, μ) be an indexed graph with multiplicities as in 7.2 and suppose that A has no loops and no multiple edges.

(1) Definition. *The* **complete indexed graph** $K(A, i_K, \mu_K)$ *with multiplicities obtained from* (A, i, μ) *consists of:*

(1) *a complete graph $K(VA)$ on VA,*

(2) *an indexing $i_K : EK(VA) \longrightarrow \mathbb{Z}^+$ such that:*

   (i) $i_K \mid_{EA} = i_A$, *and*

   (ii) *for every $e \in EK(VA) - EA$, $i_K(e)$ is chosen so that $K(A, i_K, \mu_K)$ satisfies the property (U), and*

(3) *a multiplicity $\mu_K : EK(VA) \longrightarrow \mathbb{Z}^+$ such that:*

   (i) $\mu_K \mid_{EA} = \mu_A$, *and*

   (ii) *for every $e \in EK(VA) - EA$, $\mu_K(e) = 1$.*

(2) The indexed graph $K(A, i_K, \mu_K)$ is called the *completion* of (A, i, μ) and will be denoted $K(A, i, \mu)$. We observe that $K(A, i, \mu)$ has no multiple edges.

(3) Theorem (Unique completion). *Let (A, i, μ) be an indexed graph with multiplicities and suppose that A has no loops and no multiple edges. Suppose that (A, i, μ) satisfies (F), (U), (RPE). Then there is a unique canonical completion $K(A, i, \mu)$ of (A, i, μ) satisfying (U) and (RPE).*

Proof. Let $K(VA)$ be the complete graph on VA. For every edge $e \in EK(VA) - EA$, choose a path γ_e in A from $b = \partial_1 e \in VA$ to $a = \partial_0 e \in VA$.

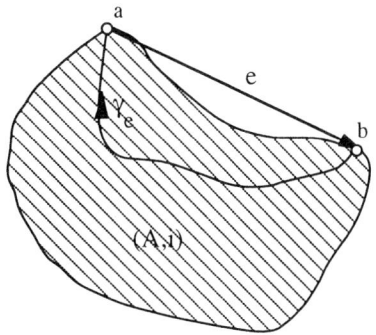

In order to ensure unimodularity of $K(A, i, \mu)$ we need to choose indices $i_K(e)$ and $i_K(\overline{e})$ so that

$$\Delta_A(\gamma_e) = \frac{\Delta_A a}{\Delta_A b} = \frac{i_K(e)}{i_K(\overline{e})}$$

for every $e \in EK(VA) - EA$.

Taking the reduced form of the rational numbers $\Delta_A(\gamma_e)$ ensures a unique choice of the indices $i_K(e)$ and $i_K(\overline{e})$, that is; suppose that $\Delta(\gamma_e) = \dfrac{s_e}{t_e}$ where $gcd(s_e, t_e) = 1$. For every $e \in EK(VA) - EA$, we take $i_K(e) = s_e$ and $i_K(\overline{e}) = t_e$.

Moreover, for every $e \in EK(VA) - EA$ we take $\mu_K(e) = \mu_K(\overline{e}) = 1$. Then $K(A, i, \mu)$ is unique and satisfies (U) and (RPE). □

(4) Corollary. *Let (A, i) be an indexed graph satisfying (F), (U), (NL), and let $(\underline{A}, \underline{i}, \underline{\mu})$ be its canonical simplification. Then there is a unique canonical completion $K(\underline{A}, \underline{i}, \underline{\mu})$ of $(\underline{A}, \underline{i}, \underline{\mu})$.*

Proof. The canonical simplification $(\underline{A}, \underline{i}, \underline{\mu})$ of (A, i) has no multiple edges and satisfies (RPE). □

8.3 Suspension

In this section, we show that *every ramified edge* in a unimodular complete indexed graph with relatively prime edges is contained in an arithmetic bridge.

(1) Let β be a geometric bridge in a graph A. Let $\overrightarrow{\beta}$ denote the graph '$\beta \cup \partial_0\beta \cup \partial_1\beta$', where

$$\partial_0\beta = \{v \in A \mid v = \partial_0 e \text{ for some } e \in \beta\}$$

$$\partial_1\beta = \{v \in A \mid v = \partial_1 e \text{ for some } e \in \beta\}.$$

(2) Let K_n denote the complete graph on n vertices. Let β be a geometric bridge in K_n. The graph $\overrightarrow{\beta}$ contains all the vertices of K_n. Moreover, β partitions the vertices of K_n into two components K_s and K_t, where $s + t = n$; in fact, K_s and K_t are complete graphs on s and t vertices respectively. Thus we can view $\overrightarrow{\beta}$ as a complete bipartite graph $K_{s,t}$, where $s + t = n$.

(3) **Lemma.** *Let K_n denote the complete graph on n vertices. Then K_n has $2^n - 2$ (oriented) geometric bridges.*

Proof. The number of geometric bridges in K_n is the number of ways of partitioning n objects into two distinct cells; there are $2^{n-1} - 1$ such ways. Taking into account the orientation, there are $2(2^{n-1} - 1) = 2^n - 2$ oriented geometric bridges. □

(4) **Definition.** *A graph A is a one-point suspension with vertex v of a subgraph $B \subsetneq A$ if*

  (1) $VA = VB \coprod \{v\}$ *for some unique $v \in VA$, and*
  (2) *the map*

$$E_0^A(v) \longrightarrow VB$$

$$e \mapsto \partial_1 e$$

is a bijection.

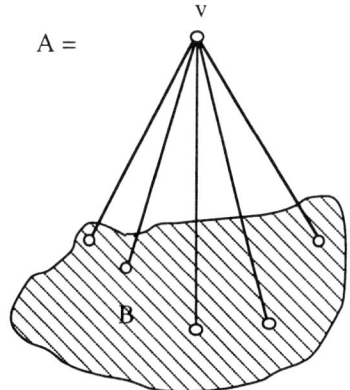

A =

(5) Remark. Let K_j denote the complete graph on j vertices. For each $j = 2, 3, \ldots$, K_j is obtained by one-point suspension of K_{j-1}:

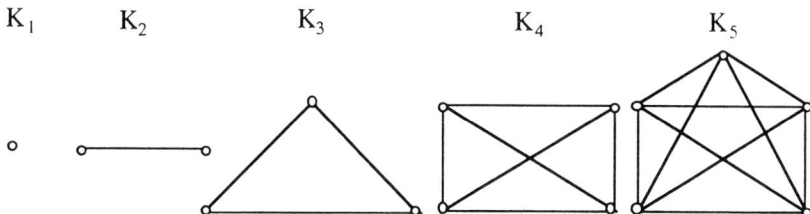

$K_1 \quad K_2 \quad K_3 \quad K_4 \quad K_5$

(6) Theorem (Suspension). Let (K_n, i) be a complete graph on n vertices with an indexing satisfying (U), and (RPE). Let (K_{n-1}, i) be a complete subgraph of (K_n, i) on $(n-1)$ vertices. If (K_{n-1}, i) has an arithmetic bridge β, then (K_n, i) has an arithmetic bridge α containing β.

Proof. Suppose that (K_{n-1}, i) has arithmetic bridge β of ramification factor $d > 1$. The graph K_n is a one-point suspension of K_{n-1} with vertex v:

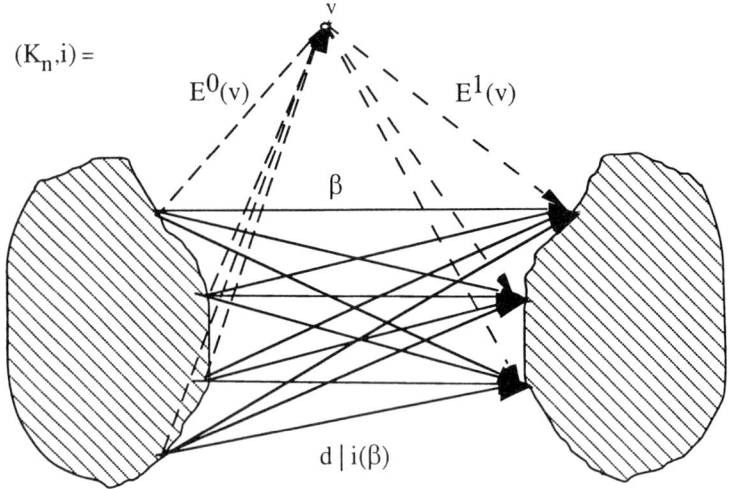

Consider in (K_n, i) the sets

$$E^0(v) = \{e \in EA \mid \partial_0 e = \partial_0 b \text{ for some } b \in \beta \text{ and } \partial_1 e = v\}$$

$$E^1(v) = \{e \in EA \mid \partial_0 e = v \text{ and } \partial_1 e = \partial_1 b \text{ for some } b \in \beta \}.$$

We claim that either $\beta \cup E^0(v)$ or $\beta \cup E^1(v)$ is an arithmetic bridge for (K_n, i); that is, there is a $d_0 > 1$ such that $d_0 \mid i(e)$ for all $e \in \beta \cup E^0(v)$, or $d_0 \mid i(e)$ for all $e \in \beta \cup E^1(v)$.

Suppose that $\beta \cup E^1(v)$ is *not* an arithmetic bridge for (K_n, i). Then there exists $f \in E^1(v)$ such that there is no prime p dividing $i(f)$ and $i(e_1)$, $i(e_2)$,... and $i(e_n)$, where $\beta = \{e_1, \ldots e_n\}$.

Choose $e \in E^0(v)$. Since $\vec{\beta}$ is a complete bipartite graph, there exists $b \in \beta$ such that $\partial_0(b) = \partial_0(e)$ and $\partial_1(b) = \partial_1(f)$:

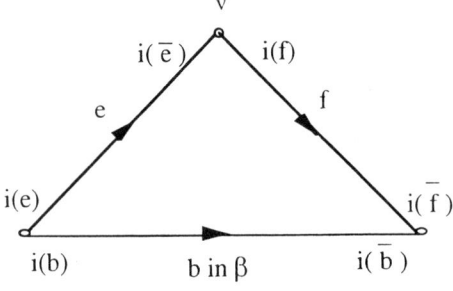

Fix a prime p dividing $i(\beta)$; that is, $p \mid i(b)$ for every $b \in \beta$ (we can find such a p since β has ramification factor $d > 1$). Thus $p \mid i(b)$. We claim that $p \mid i(e)$.

By (U), we have:

$$\frac{i(\overline{e})}{i(e)} \cdot \frac{i(\overline{f})}{i(f)} \cdot \frac{i(b)}{i(\overline{b})} = 1.$$

By (RPE), $p \nmid i(\overline{b})$. By assumption, $p \nmid i(f)$. By (U), we must have $p \mid i(e)$.

Moreover, this is independant of the choice of $e \in E^0(v)$; for any $e \in E^0(v)$ we can find $b \in \beta$ such that $\partial_0(b) = \partial_0(e)$ and $\partial_1(b) = \partial_1(f)$ since $\overrightarrow{\beta}$ is a complete bipartite graph $K_{s,t}$.

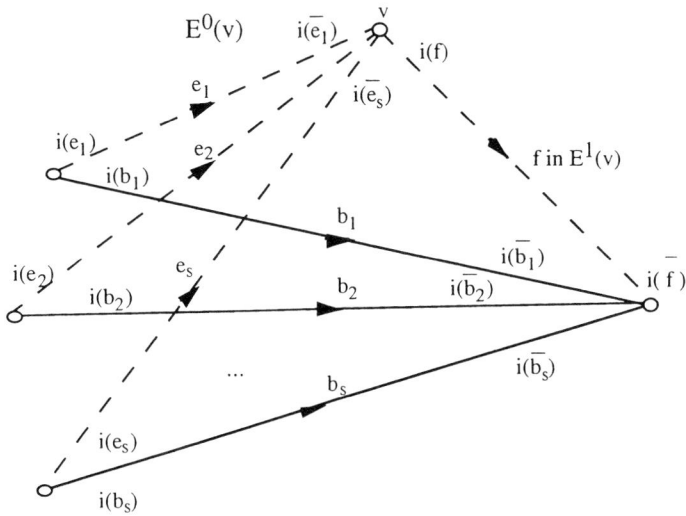

We assume that $p \nmid i(f)$, and by (RPE), $p \nmid i(\overline{b_j})$ for every $b_j \in \beta$, thus we conclude that $p \mid i(e)$ for every $e \in E^0(v)$; that is, $\beta \cup E^0(v)$ is an arithmetic bridge α for (K_n, i) of ramification factor $d_0 = p$.

□

(7) Corollary. *Let (A, i, μ) be an indexed graph with multiplicities and suppose that A has no loops and no multiple edges. Suppose that (A, i, μ) satisfies (F), (U), (RPE). Let*

$K(A, i, \mu)$ be the canonical completion of (A, i, μ). Then every ramified edge of $K(A, i, \mu)$ is contained in an arithmetic bridge in $K(A, i, \mu)$.

Proof. Choose a ramified edge e of $K(A, i, \mu)$. Since $K(A)$ is a complete graph, $K(A)$ is constructed from iterated one-point suspensions $K_1 = \partial_0 e$, $K_2 = e$, ..., $K_n = K(A)$. Moreover, e is a geometric bridge for K_2, in fact an arithmetic bridge for (K_2, i, μ). By the suspension theorem 8.3.6, every one-point suspension $(K_2, i, \mu) = e$, ..., $(K_n, i, \mu) = K(A, i, \mu)$ has an arithmetic bridge containing e. □

(8) Corollary. *Let (A, i) be an indexed graph satisfying (F), (U), (NL), and let $(\underline{A}, \underline{i}, \underline{\mu})$ be its canonical simplification. Let $K(\underline{A}, \underline{i}, \underline{\mu})$ be the unique canonical completion of $(\underline{A}, \underline{i}, \underline{\mu})$. Then every ramified edge e of $K(\underline{A}, \underline{i}, \underline{\mu})$ is contained in an arithmetic bridge in $K(\underline{A}, \underline{i}, \underline{\mu})$.*

Proof. The canonical simplification $(\underline{A}, \underline{i}, \underline{\mu})$ has no multiple edges and automatically satisfies (RPE).□

8.4 Restriction

In this section, we show that by choosing the arithmetic bridge β in $K(\underline{A}, \underline{i}, \underline{\mu})$ such that $\beta \cap \underline{A} \neq \emptyset$, it follows that $\beta \cap \underline{A}$ contains an arithmetic bridge α for $(\underline{A}, \underline{i}, \underline{\mu})$.

(1) Theorem (Existence of geometric bridges). *Let A be a finite connected graph with no loops. Suppose that A contains a (p, q)-geometric bridge. Then A contains a $(1, 1)$-geometric bridge.*

Proof. Fix $A_k \in \partial_0 \beta$. Let $E_0(A_k, \beta) = \{e \in \beta \mid \partial_0 e \in A_k\}$:

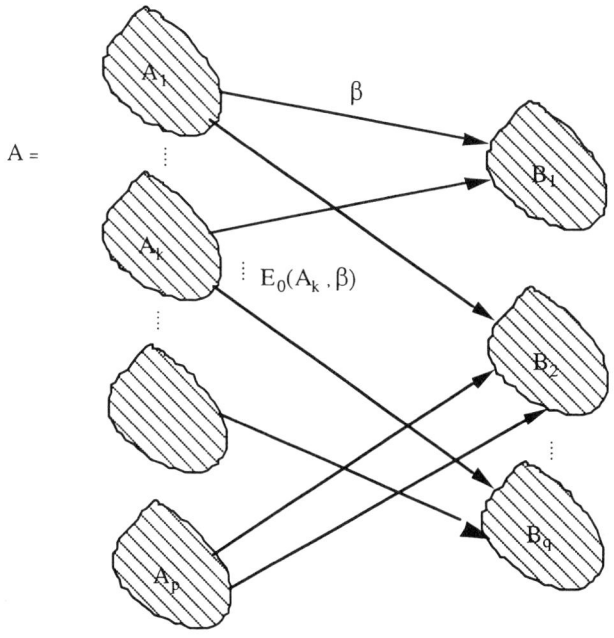

If
$$A - \{A_k \cup E_0(A_k, \beta) \cup \overline{E_0(A_k, \beta)}\}$$

is connected, then we are done; $E_0(A_k, \beta)$ is a $(1,1)$-geometric bridge from A_k to $A - \{A_k \cup E_0(A_k, \beta) \cup \overline{E_0(A_k, \beta)}\}$:

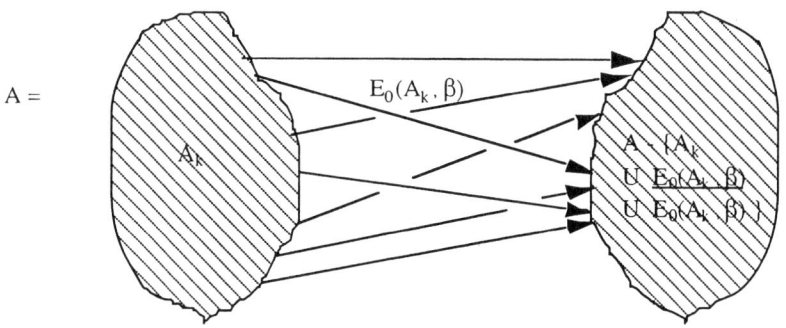

Suppose, conversely, that $A - \{A_k \cup E_0(A_k, \beta) \cup \overline{E_0(A_k, \beta)}\}$ has connected components

$C_1, \ldots, C_t$:

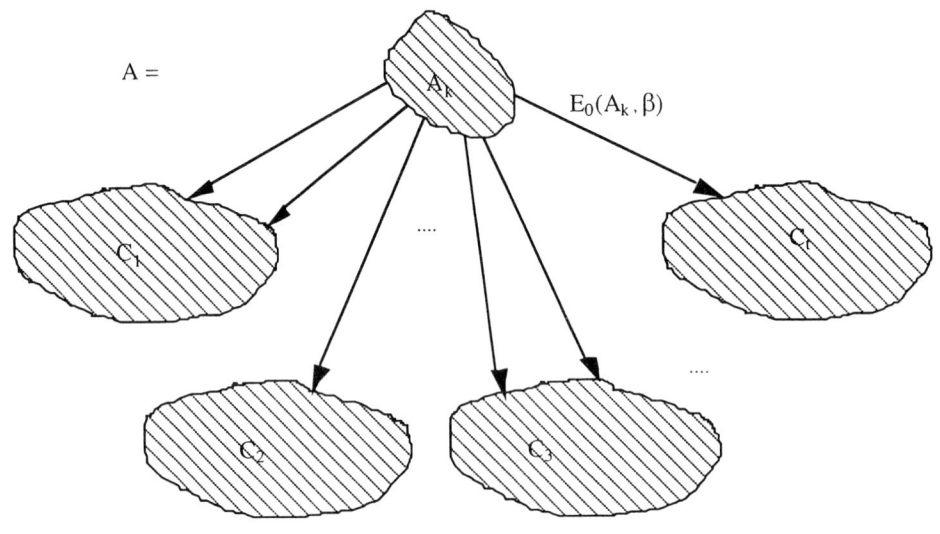

Fix a connected component C_l amongst the $C_1, \ldots, C_t$. Let

$$E_0(A_k, \beta, C_l) = \{e \in E_0(A_k, \beta) \mid \partial_0 e \in A_k, \partial_1 e \in C_l\}.$$

Then $E_0(A_k, \beta, C_l)$ is a $(1,1)$-geometric bridge from

$$A - \{C_l \cup E_0(A_k, \beta, C_l) \cup \overline{E_0(A_k, \beta, C_l)}\}$$

to C_l:

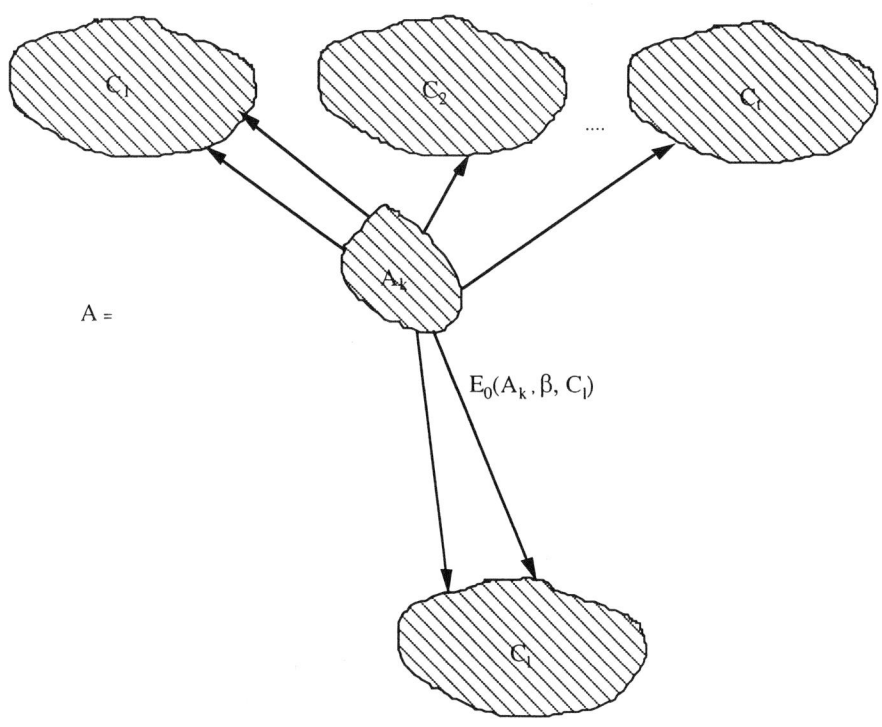

(2) Theorem (Existence of arithmetic bridges I). *Let (A, i, μ) be an edge-indexed graph with multiplicities. Suppose that (A, i, μ) satisfies the conditions (F), (U), (MIN), (NDR), (RPE), has no loops and no multiple edges. Then (A, i, μ) contains a good separating edge, or an arithmetic bridge with $n \geq 2$ edges.*

Proof. We form $K(A, i, \mu)$, the unique canonical completion of (A, i, μ). By Corollary 8.3.7, every edge of $K(A, i, \mu)$ is contained in an arithmetic bridge in $K(A, i, \mu)$. Choose an arithmetic bridge α in $K(A, i, \mu)$ such that $\alpha \cap (A, i, \mu) \neq \varnothing$. Take the restriction of

$K(A, i, \mu)$ to (A, i, μ):

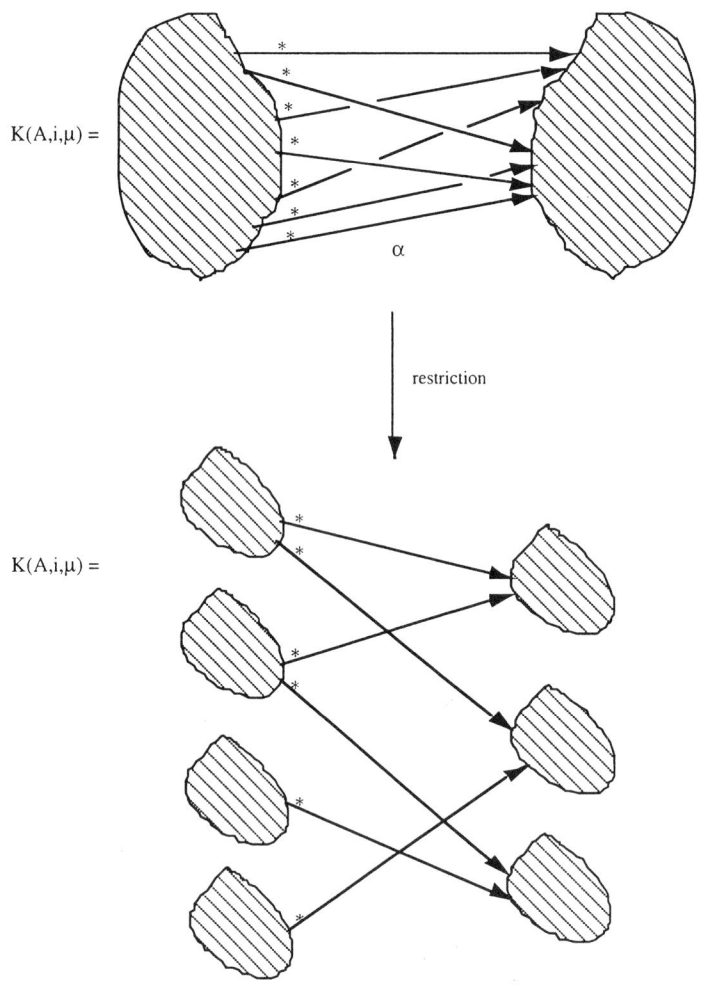

The restriction of the arithmetic bridge α in $K(A, i, \mu)$ to (A, i, μ) is a (p, q)-arithmetic bridge β for (A, i, μ). By Theorem 8.4.1 (Existence of geometric bridges), β contains a $(1,1)$-geometric bridge β_0 which is arithmetic for (A, i, μ).

If β_0 is a single separating edge of the form:

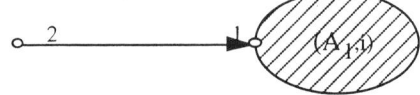

with $\pi_1(A_1) \neq 1$, then β_0 is not a good separating edge for (A, i, μ). By $(NDR)_{min}$, (A_1, i, μ) must contain another ramified edge e. By corollary 8.3.7 to the suspension theorem, e is contained in an arithmetic bridge β_e^K in $K(A_1, i, \mu)$. We take the restriction $\beta_e^{A_1}$ of β_e^K in $K(A_1, i, \mu)$. By theorem 8.4.1 (Existence of geometric bridges), $\beta_e^{A_1}$ contains a (1,1)-geometric bridge $\beta_0^{A_1}$ which is arithmetic for (A_1, i, μ).

If $\beta_0^{A_1}$ has ≥ 2 edges, then we are done. If $|\beta_0^{A_1}| = 1$, then (A, i, μ) is of the form:

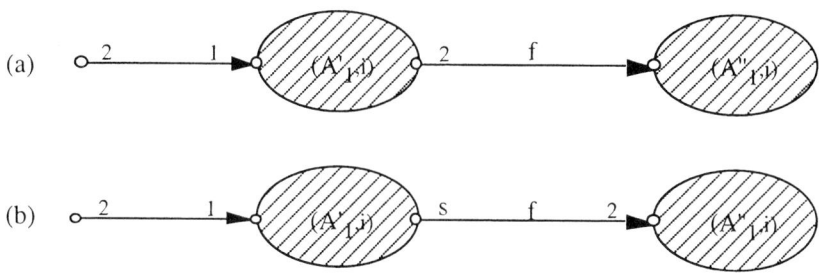

In (a), f is a good separating edge for (A, i, μ). Suppose we are in case (b). If $A_1'' \neq \{\circ\}$, then f is a good separating edge for (A, i, μ). Suppose $A_1'' = \{\circ\}$. If $s \geq 2$ then f is a good separating edge for (A, i, μ). Suppose $s = 1$:

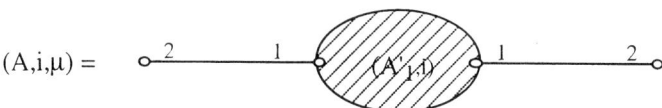

Then A_1' must contain a ramified edge h or (A, i, μ) violates $(NDR)_{min}$. By Corollary 8.3.7, h is contained in an arithmetic bridge β_h^K in $K(A_1', i, \mu)$. We take the restriction $\beta_e^{A_1'}$ of β_h^K in $K(A_1', i, \mu)$. By Theorem 8.4.1 (Existence of geometric bridges), $\beta_e^{A_1'}$ contains a (1,1)-geometric bridge $\beta_0^{A_1'}$ which is arithmetic for (A_1', i, μ).

If $\beta_0^{A_1'}$ has ≥ 2 edges, then we are done. If $|\beta_0^{A_1'}| = 1$, then (A, i, μ) is of the form:

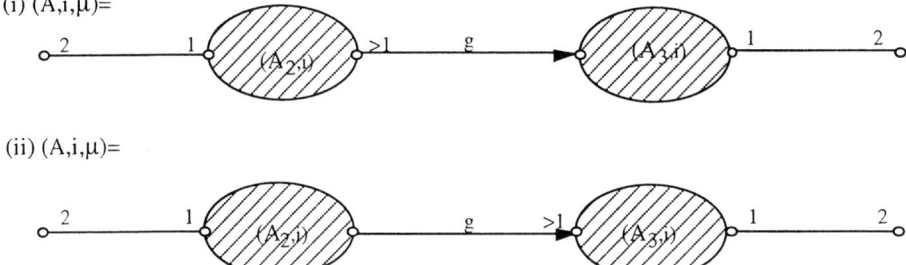

and in both (i) and (ii), g is a good separating edge for (A, i, μ).□

(3) Remark. *The theorem is true if we drop the condition (MIN) and then $(NDR)_{min}$ takes the form (NDR) as described in 8.1.2.*

(4) Corollary (Existence of arithmetic bridges in the core). *Let (A, i) be an indexed graph satisfying (F), (U), (NDR), (MIN), (RPE) and (NL). Then (A, i) contains a good separating edge, or an arithmetic bridge with $n \geq 2$ edges.*

Proof. Let $(\underline{A}, \underline{i}, \mu)$ be the canonical simplification of (A, i). Then $(\underline{A}, \underline{i}, \mu)$ has no loops, no multiple edges and satisfies (F), (U), and (RPE). We claim that we can also assume that $(\underline{A}, \underline{i}, \mu)$ satisfies (NDR) and (MIN).

If (A, i) satisfies $(NDR)_{min}$ and (A, i) is strongly ramified, then $(\underline{A}, \underline{i}, \mu)$ satisfies $(NDR)_{min}$. If (A, i) is weakly ramified but satisfies $(NDR)_{min}$, and $(\underline{A}, \underline{i}, \mu)$ violates $(NDR)_{min}$, then the only possibilities for (A, i) are:

(a) (A, i) is of the form:

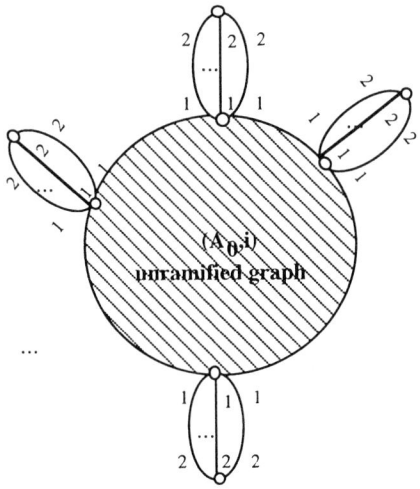

where (A_0, i) is an unramified graph. Then $(\underline{A}, \underline{i}, \underline{\mu})$ violates $(NDR)_{min}$, that is $(\underline{A}, \underline{i}, \underline{\mu})$ satisfies $(DR)_{min}$, but at least one of the subgraphs, a 'cage'

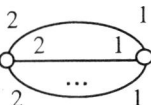

must have at least 2 edges, and hence is an arithmetic bridge for (A, i), and we are done.

(b) (A,i) is of the form:

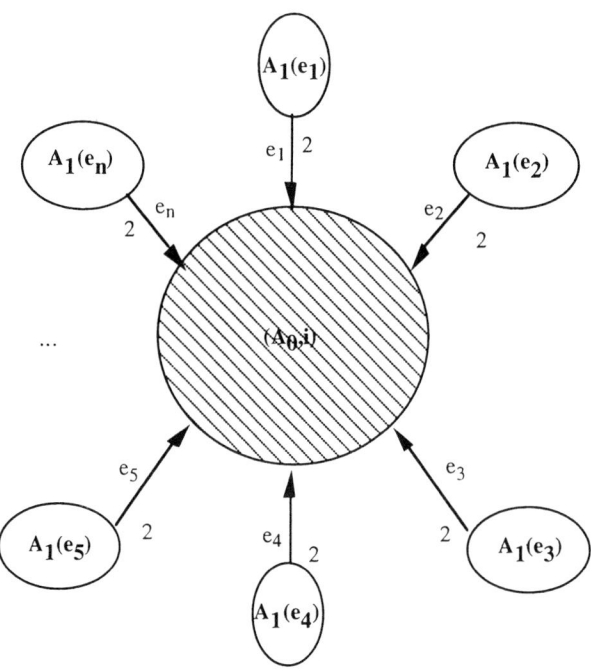

where (A_0, i) is an unramified graph and $(A_k(e_k), i)$, $k = 1, \ldots, n$ are bouquets of unramified cages. Then $(\underline{A}, \underline{i}, \underline{\mu})$ violates (MIN) since each $(A_k(e_k), i)$, $k = 1, \ldots, n$ becomes an unramified tree in $(\underline{A}, \underline{i}, \underline{\mu})$, and $(\underline{A}, \underline{i}, \underline{\mu})$ violates (NDR). However, we must have $(A_j(e_j), i) \neq \varnothing$ for some $j \in \{1, \ldots, n\}$ or (A, i) violates $(NDR)_{min}$. If $(A_j(e_j), i) \neq \varnothing$, then the ramified edge e_j is a good separating edge for (A, i), and we are done.

(c) (A, i) is of the form:

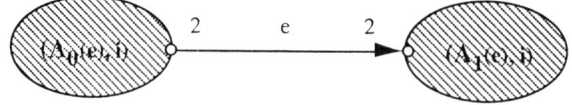

where $(A_k(e), i)$ $k = 0, 1$ are bouquets of unramified cages. Then in $(\underline{A}, \underline{i}, \underline{\mu})$, $(A_k(e), i)$, $k = 0, 1$ become unramified trees. Thus $(\underline{A}, \underline{i}, \underline{\mu})$ violates (MIN) and (NDR), but e is a good separating edge for (A, i) and we are done.

In addition to the above cases, it may happen that (A, i) satisfies (MIN), but $(\underline{A}, \underline{i}, \underline{\mu})$ violates (MIN). In this case, (A, i) must be of the form:

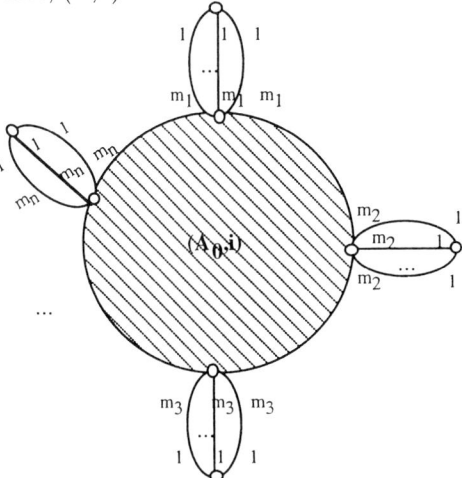

where $m_j \geq 1$, $j = 1, \ldots, n$. Since (A, i) satisfies (MIN), each cage:

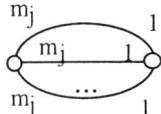

must contain at least 2 edges. If $m_j > 1$ for some $j = 1, \ldots, n$, then the corresponding cage is an arithmetic bridge for (A, i), and we are done.

If $m_j = 1$ for each $j = 1, \ldots, n$, then (A_0, i) must contain another ramified edge, or (A, i) violates $(NDR)_{min}$. If we remove all unramified cages from (A, i), then the canonical simplification of the resulting edge-indexed graph (A_0, i) will satisfy (MIN), and if we can prove the corollary for (A_0, i), then a good separating edge or arithmetic bridge for (A_0, i) is a good separating edge or arithmetic bridge for (A, i), and we are done.

Hence we can assume that $(\underline{A}, \underline{i}, \underline{\mu})$ has no loops, no multiple edges, satisfies (F), (U), (RPE), $(NDR)_{min}$ and (MIN).

Thus by Theorem 8.4.2 (Existence of arithmetic bridges), $(\underline{A}, \underline{i}, \underline{\mu})$ contains contains an arithmetic bridge with $n \geq 2$ edges, or a 'good' separating edge.

We argue that an arithmetic bridge or a good separating edge for $(\underline{A}, \underline{i}, \underline{\mu})$ implies the existence of a an arithmetic bridge or a good separating edge for (A, i).

Suppose that $(\underline{A}, \underline{i}, \underline{\mu})$ contains an arithmetic bridge $\underline{\beta}$ of ramification factor $d \geq 2$. The following argument holds when $\underline{\beta}$ consists of $n \geq 2$ edges, or when $\underline{\beta}$ is a single edge; that is, a 'good' separating-edge for $(\underline{A}, \underline{i}, \underline{\mu})$.

We recall that since (A, i) satisfies (RPE), $(\underline{A}, \underline{i}, \underline{\mu})$ is obtained from (A, i) by replacing each multiple edge $E^A(a, b)$ of (A, i) with a single edge (a, b) with multiplicity $\underline{\mu}$, where $\underline{\mu} = |E^A(a, b)|$.

Let $q : A \longrightarrow \underline{A}$ be the natural projection (see 7.1). Let $e \in \underline{\beta} \subset E\underline{A}$. Since $\underline{\beta}$ has ramification factor $d \geq 2$, we have $d | i(f)$ for every $f \in \underline{\beta}$. The set

$$\beta = \{e \in q^{-1}(f) \mid f \in \underline{\beta} \subset E\underline{A}\}$$

is a geometric bridge for A, since $\underline{\beta}$ is a geometric bridge for $\underline{A}$. Moreover, $d | i(e)$ for every $e \in \beta \subset EA$, and so β is an arithmetic bridge for (A, i) of ramification factor d. $\square$

(5) Remark. *This corollary is true if we drop condition (MIN) (see Remark 8.4.3).*

A corollary of 8.4.2 and 8.4.4 is:

(6) Corollary (Existence of arithmetic bridges II). *Let (A, i) be an edge-indexed graph. Suppose that (A, i) satisfies (U), (F), (NDR), (MIN), (RPE). Suppose that (A, i) has no loops and no multiple edges. Let $e \in EA$ be a ramified edge. If e is not separating, then e is contained in an arithmetic bridge with $n \geq 2$ edges*

Corollary 8.4.6 is true if we drop conditions (MIN) and (NDR). This can be seen as follows. If (A, i) is finite and *discretely ramified*, then $((A, i)$ automatically satisfies (U) and (RPE), and) every ramified edge is separating, so Corollary 8.4.6 is true. (If (A, i) satisfies (NDR) but not (MIN), then by Remarks 8.4.3 and 8.4.5, Theorem 8.4.2 and Corollary 8.4.4 are also true.) Hence we obtain:

(7) Corollary (Existence of arithmetic bridges III). *Let (A, i) be an edge-indexed graph. Suppose that (A, i) satisfies (U), (F), and (RPE). Suppose that (A, i) has no loops and no multiple edges. Let $e \in EA$ be a ramified edge. If e is not separating, then e is contained in an arithmetic bridge with $n \geq 2$ edges*

Finally, as a corollary of 8.4.4, we obtain Theorem 8.1.4:

(8) Theorem (Existence of arithmetic bridges IV). *Let (A, i) be an edge-indexed graph. Suppose that (A, i) satisfies (U), (F), (NDR), (MIN), (RPE) and (UL). Then (A, i) contains either an arithmetic bridge with $n \geq 2$ edges, or a 'good' separating edge.*

In conclusion, by Theorems 4.5.2, 4.13.1, 5.27, 6.6, we have

Theorem (Existence of non-uniform coverings). *Let (A, i) be an edge-indexed graph. Suppose that (A, i) satisfies (U), (F), (NDR), (MIN). If (A, i) contains an arithmetic bridge with $n \geq 2$ edges, an edge whose indices have a common factor $n > 1$, a 'good' separating edge or a ramified loop, then (A, i) has a non-uniform covering.*

By Theorem 8.1.4, if all loops of (A, i) are unramified, then under the assumptions (U), (F), (NDR), (MIN), and (RPE) (A, i) must contain an arithmetic bridge, or a good separating edge. Hence the proof of Conjecture 3.4.13 (Existence of non-uniform lattices on uniform trees) is complete.

BIBLIOGRAPHY

[B] Bass H, *Covering theory for graphs of groups*, Journal of Pure and Applied Algebra **89** (1993).

[BCR] Bass, H, Carbone L, and Rosenberg, G, *The Existence Theorem for tree lattices*, Appendix B, '*Tree Lattices*' by Hyman Bass and Alex Lubotzky (2000), Birkhauser, Boston.

[Bo1] Borel, A, *Introduction aux groupes arithmetiques*, Herman, Paris (1990).

[Bo2] Borel, A, *Compact Clifford-Klein forms of symmetric spaces*, Topology **2** (1963).

[BH] Borel, A, Harder, G, *Existence of discrete cocompact subgroups of reductive groups over local fields*, J. Reine Angew. Math **298** (1978).

[BK] Bass H and Kulkarni R, *Uniform tree lattices*, Journal of the Amer Math Society **3 (4)** (1990).

[BL] Bass H and Lubotzky A, *Tree lattices* (2000), Birkhauser, Boston.

[BM] Burger, M and Mozes, S, *CAT(-1) Spaces, divergence groups and their commensurators*, Preprint.

[BT] Bass H and Tits J, *A Discreteness Criterion for certain tree automorphism groups*, Appendix A, '*Tree Lattices*' by Hyman Bass and Alex Lubotzky (2000), Birkhauser, Boston.

[C1] Carbone L, *Non-uniform lattices on uniform trees*, PhD. Thesis, Columbia University (1997).

[C2] Carbone, L, *Non-minimal tree actions and the existence of non-uniform tree lattices*, (in preparation) (2000).

[C3] Carbone, L, *Constructing tree lattices*, Algebras and Combinatorics. An International Congress, ICAC '97, Hong Kong (ed K.P. Shum and E. Taft) (**pp 63-97**) (1999), Springer.

[CC] Carbone, L and Clark, D, *Lattices on parabolic trees*, Preprint (2000).

[CR1] Carbone, L and Rosenberg, G, *Lattices on non-uniform trees*, Preprint (2000).

[CR2] Carbone, L and Rosenberg, G, *Infinite towers of tree lattices*, Preprint (2000).

[CR3] Carbone, L and Rosenberg, G, *Infinite towers of non-uniform tree lattices*, (in preparation) (2000).

[K] Kulkarni R, *Lattices on trees, automorphisms of graphs, free groups and surfaces*, Preprint (1993).

[IL] Levich, Inga, *PhD thesis*, Hebrew University of Jerusalem (1996).

[L1] Lubotzky A, *Tree lattices and lattices in Lie groups*, in *Combinatorial and Geometric Group Theory*, ed A. Duncan, N. Gilbert and J. Howie, LMS Lecture Note Series 204 (**pp 217-232**) (1995), Cambridge University Press.

[L2] Lubotzky A, *Lattices in rank one Lie groups over local fields*, Geometric and Functional Analysis **4** (**pp 405-431**) (1991).

[L3] Lubotzky A, *Lattices of minimal covolume in SL_2*, Jour of the AMS **3** (**pp 961-975**) (1990).

[LMZ] Lubotzky, A, Mozes, S and Zimmer, R, *Superrigidity for the commensurability group of tree lattices*, Comment. Math. Helvetici **69** (**pp 523-548**) (1994).

[Ma] Margulis, G, *Discrete subgroups of semi-simple Lie groups*, Springer-Verlag (1991).

[Mo] Mozes, S, *The congruence subgroup problem for uniform tree lattices*, Preprint.

[Mo2] Mozes, S, *Private communication* (1997).

[R] Rosenberg, G, *Infinite towers of uniform tree lattices*, PhD. Thesis, Columbia University (2000).

[S] Serre, J.P, *Trees (Translated from the French by John Stilwell)* (1980), Springer-Verlag, Berlin Heidelberg.

[Ta] Tamagawa, T, *On discrete subgroups of p-adic algebraic groups*, Arithmetical Algebraic Geometry (ed O. F. G. Schilling) (1965), Harper and Row.

[Ti]   Tits, J, *Sur le groupe des automorphisms d'un arbe*, Essays on topology and related topics: Memories dedies a George de Rham (1970), Springer.
[VH]   Valette, A and de la Harpe, P, *La propriete (T) de Kazhdan pour les groupes localement compacts (avec un appendice de Marc Burger)*, Asterisque **175** (1989).
[YL]   Liu, YS, *Density of the commensurability group of uniform tree lattices*, Journal of Algebra **165** (1994).

Department of Mathematics, Harvard University, Science Center 325, 1 Oxford St, Cambridge MA 02138

e-mail lisa@ math.harvard.edu

Editors

This journal is designed particularly for long research papers, normally at least 80 pages in length, and groups of cognate papers in pure and applied mathematics. Papers intended for publication in the *Memoirs* should be addressed to one of the following editors. In principle the Memoirs welcomes electronic submissions, and some of the editors, those whose names appear below with an asterisk (*), have indicated that they prefer them. However, editors reserve the right to request hard copies after papers have been submitted electronically. Authors are advised to make preliminary email inquiries to editors about whether they are likely to be able to handle submissions in a particular electronic form.

Algebra to CHARLES CURTIS, Department of Mathematics, University of Oregon, Eugene, OR 97403-1222 email: `cwc@darkwing.uoregon.edu`

Algebraic geometry and commutative algebra to LAWRENCE EIN, Department of Mathematics, University of Illinois, 851 S. Morgan (M/C 249), Chicago, IL 60607-7045; email: `ein@uic.edu`

Algebraic topology and cohomology of groups to STEWART PRIDDY, Department of Mathematics, Northwestern University, 2033 Sheridan Road, Evanston, IL 60208-2730; email: `priddy@math.nwu.edu`

Combinatorics and Lie theory to SERGEY FOMIN, Department of Mathematics, University of Michigan, Ann Arbor, Michigan 48109-1109; email: `fomin@math.lsa.umich.edu`

Complex analysis and complex geometry to DUONG H. PHONG, Department of Mathematics, Columbia University, 2990 Broadway, New York, NY 10027-0029; email: `dp@math.columbia.edu`

*__Differential geometry and global analysis__ to LISA C. JEFFREY, Department of Mathematics, University of Toronto, 100 St. George St., Toronto, ON Canada M5S 3G3; email: `jeffrey@math.toronto.edu`

*__Dynamical systems and ergodic theory__ to ROBERT F. WILLIAMS, Department of Mathematics, University of Texas, Austin, Texas 78712-1082; email: `bob@math.utexas.edu`

Functional analysis and operator algebras to BRUCE E. BLACKADAR, Department of Mathematics, University of Nevada, Reno, NV 89557; email: `bruceb@math.unr.edu`

Geometric topology, knot theory and hyperbolic geometry to ABIGAIL A. THOMPSON, Department of Mathematics, University of California, Davis, Davis, CA 95616-5224; email: `thompson@math.ucdavis.edu`

Harmonic analysis, representation theory, and Lie theory to ROBERT J. STANTON, Department of Mathematics, The Ohio State University, 231 West 18th Avenue, Columbus, OH 43210-1174; email: `stanton@math.ohio-state.edu`

*__Logic__ to THEODORE SLAMAN, Department of Mathematics, University of California, Berkeley, CA 94720-3840; email: `slaman@math.berkeley.edu`

Number theory to MICHAEL J. LARSEN, Department of Mathematics, Indiana University, Bloomington, IN 47405; email: `larsen@math.indiana.edu`

*__Ordinary differential equations, partial differential equations, and applied mathematics__ to PETER W. BATES, Department of Mathematics, Brigham Young University, 292 TMCB, Provo, UT 84602-1001; email: `peter@math.byu.edu`

*__Partial differential equations and applied mathematics__ to BARBARA LEE KEYFITZ, Department of Mathematics, University of Houston, 4800 Calhoun Road, Houston, TX 77204-3476; email: `keyfitz@uh.edu`

*__Probability and statistics__ to KRZYSZTOF BURDZY, Department of Mathematics, University of Washington, Box 354350, Seattle, Washington 98195-4350; email: `burdzy@math.washington.edu`

*__Real and harmonic analysis and geometric partial differential equations__ to WILLIAM BECKNER, Department of Mathematics, University of Texas, Austin, TX 78712-1082; email: `beckner@math.utexas.edu`

All other communications to the editors should be addressed to the Managing Editor, WILLIAM BECKNER, Department of Mathematics, University of Texas, Austin, TX 78712-1082; email: `beckner@math.utexas.edu`.

Editorial Information

To be published in the *Memoirs*, a paper must be correct, new, nontrivial, and significant. Further, it must be well written and of interest to a substantial number of mathematicians. Piecemeal results, such as an inconclusive step toward an unproved major theorem or a minor variation on a known result, are in general not acceptable for publication. Papers appearing in *Memoirs* are generally longer than those appearing in *Transactions*, which shares the same editorial committee.

As of March 31, 2001, the backlog for this journal was approximately 6 volumes. This estimate is the result of dividing the number of manuscripts for this journal in the Providence office that have not yet gone to the printer on the above date by the average number of monographs per volume over the previous twelve months, reduced by the number of volumes published in four months (the time necessary for preparing a volume for the printer). (There are 6 volumes per year, each containing at least 4 numbers.)

A Consent to Publish and Copyright Agreement is required before a paper will be published in the *Memoirs*. After a paper is accepted for publication, the Providence office will send a Consent to Publish and Copyright Agreement to all authors of the paper. By submitting a paper to the *Memoirs*, authors certify that the results have not been submitted to nor are they under consideration for publication by another journal, conference proceedings, or similar publication.

Information for Authors

Memoirs are printed from camera copy fully prepared by the author. This means that the finished book will look exactly like the copy submitted.

The paper must contain a *descriptive title* and an *abstract* that summarizes the article in language suitable for workers in the general field (algebra, analysis, etc.). The *descriptive title* should be short, but informative; useless or vague phrases such as "some remarks about" or "concerning" should be avoided. The *abstract* should be at least one complete sentence, and at most 300 words. Included with the footnotes to the paper should be the 2000 *Mathematics Subject Classification* representing the primary and secondary subjects of the article. The classifications are accessible from www.ams.org/msc/. The list of classifications is also available in print starting with the 1999 annual index of *Mathematical Reviews*. The Mathematics Subject Classification footnote may be followed by a list of *key words and phrases* describing the subject matter of the article and taken from it. Journal abbreviations used in bibliographies are listed in the latest *Mathematical Reviews* annual index. The series abbreviations are also accessible from www.ams.org/publications/. To help in preparing and verifying references, the AMS offers MR Lookup, a Reference Tool for Linking, at www.ams.org/mrlookup/. When the manuscript is submitted, authors should supply the editor with electronic addresses if available. These will be printed after the postal address at the end of the article.

Electronically prepared manuscripts. The AMS encourages electronically prepared manuscripts, with a strong preference for $\mathcal{AMS}$-LaTeX. To this end, the Society has prepared $\mathcal{AMS}$-LaTeX author packages for each AMS publication. Author packages include instructions for preparing electronic manuscripts, the *AMS Author Handbook*, samples, and a style file that generates the particular design specifications of that publication series. Though $\mathcal{AMS}$-LaTeX is the highly preferred format of TeX, author packages are also available in $\mathcal{AMS}$-TeX.

Authors may retrieve an author package from e-MATH starting from `www.ams.org/tex/` or via FTP to `ftp.ams.org` (login as `anonymous`, enter username as password, and type `cd pub/author-info`). The *AMS Author Handbook* and the *Instruction Manual* are available in PDF format following the author packages link from `www.ams.org/tex/`. The author package can be obtained free of charge by sending email to `pub@ams.org` (Internet) or from the Publication Division, American Mathematical Society, P.O. Box 6248, Providence, RI 02940-6248. When requesting an author package, please specify $\mathcal{AMS}$-LaTeX or $\mathcal{AMS}$-TeX, Macintosh or IBM (3.5) format, and the publication in which your paper will appear. Please be sure to include your complete mailing address.

Sending electronic files. After acceptance, the source file(s) should be sent to the Providence office (this includes any TeX source file, any graphics files, and the DVI or PostScript file).

Before sending the source file, be sure you have proofread your paper carefully. The files you send must be the EXACT files used to generate the proof copy that was accepted for publication. For all publications, authors are required to send a printed copy of their paper, which exactly matches the copy approved for publication, along with any graphics that will appear in the paper.

TeX files may be submitted by email, FTP, or on diskette. The DVI file(s) and PostScript files should be submitted only by FTP or on diskette unless they are encoded properly to submit through email. (DVI files are binary and PostScript files tend to be very large.)

Electronically prepared manuscripts can be sent via email to `pub-submit@ams.org` (Internet). The subject line of the message should include the publication code to identify it as a Memoir. TeX source files, DVI files, and PostScript files can be transferred over the Internet by FTP to the Internet node `e-math.ams.org` (130.44.1.100).

Electronic graphics. Comprehensive instructions on preparing graphics are available at `www.ams.org/jourhtml/graphics.html`. A few of the major requirements are given here.

Submit files for graphics as EPS (Encapsulated PostScript) files. This includes graphics originated via a graphics application as well as scanned photographs or other computer-generated images. If this is not possible, TIFF files are acceptable as long as they can be opened in Adobe Photoshop or Illustrator. No matter what method was used to produce the graphic, it is necessary to provide a paper copy to the AMS.

Authors using graphics packages for the creation of electronic art should also avoid the use of any lines thinner than 0.5 points in width. Many graphics packages allow the user to specify a "hairline" for a very thin line. Hairlines often look acceptable when proofed on a typical laser printer. However, when produced on a high-resolution laser imagesetter, hairlines become nearly invisible and will be lost entirely in the final printing process.

Screens should be set to values between 15% and 85%. Screens which fall outside of this range are too light or too dark to print correctly. Variations of screens within a graphic should be no less than 10%.

Inquiries. Any inquiries concerning a paper that has been accepted for publication should be sent directly to the Electronic Prepress Department, American Mathematical Society, P. O. Box 6248, Providence, RI 02940-6248.

Selected Titles in This Series

(Continued from the front of this publication)

694 **Alberto Bressan, Graziano Crasta, and Benedetto Piccoli,** Well-posedness of the Cauchy problem for $n \times n$ systems of conservation laws, 2000

693 **Doug Pickrell,** Invariant measures for unitary groups associated to Kac-Moody Lie algebras, 2000

692 **Mara D. Neusel,** Inverse invariant theory and Steenrod operations, 2000

691 **Bruce Hughes and Stratos Prassidis,** Control and relaxation over the circle, 2000

690 **Robert Rumely, Chi Fong Lau, and Robert Varley,** Existence of the sectional capacity, 2000

689 **M. A. Dickmann and F. Miraglia,** Special groups: Boolean-theoretic methods in the theory of quadratic forms, 2000

688 **Piotr Hajłasz and Pekka Koskela,** Sobolev met Poincaré, 2000

687 **Guy David and Stephen Semmes,** Uniform rectifiability and quasiminimizing sets of arbitrary codimension, 2000

686 **L. Gaunce Lewis, Jr.,** Splitting theorems for certain equivariant spectra, 2000

685 **Jean-Luc Joly, Guy Metivier, and Jeffrey Rauch,** Caustics for dissipative semilinear oscillations, 2000

684 **Harvey I. Blau, Bangteng Xu, Z. Arad, E. Fisman, V. Miloslavsky, and M. Muzychuk,** Homogeneous integral table algebras of degree three: A trilogy, 2000

683 **Serge Bouc,** Non-additive exact functors and tensor induction for Mackey functors, 2000

682 **Martin Majewski,** ational homotopical models and uniqueness, 2000

681 **David P. Blecher, Paul S. Muhly, and Vern I. Paulsen,** Categories of operator modules (Morita equivalence and projective modules, 2000

680 **Joachim Zacharias,** Continuous tensor products and Arveson's spectral C^*-algebras, 2000

679 **Y. A. Abramovich and A. K. Kitover,** Inverses of disjointness preserving operators, 2000

678 **Wilhelm Stannat,** The theory of generalized Dirichlet forms and its applications in analysis and stochastics, 1999

677 **Volodymyr V. Lyubashenko,** Squared Hopf algebras, 1999

676 **S. Strelitz,** Asymptotics for solutions of linear differential equations having turning points with applications, 1999

675 **Michael B. Marcus and Jay Rosen,** Renormalized self-intersection local times and Wick power chaos processes, 1999

674 **R. Lawther and D. M. Testerman,** A_1 subgroups of exceptional algebraic groups, 1999

673 **John Lott,** Diffeomorphisms and noncommutative analytic torsion, 1999

672 **Yael Karshon,** Periodic Hamiltonian flows on four dimensional manifolds, 1999

671 **Andrzej Rosłanowski and Saharon Shelah,** Norms on possibilities I: Forcing with trees and creatures, 1999

670 **Steve Jackson,** A computation of δ_5^1, 1999

669 **Seán Keel and James McKernan,** Rational curves on quasi-projective surfaces, 1999

668 **E. N. Dancer and P. Poláčik,** Realization of vector fields and dynamics of spatially homogeneous parabolic equations, 1999

667 **Ethan Akin,** Simplicial dynamical systems, 1999

666 **Mark Hovey and Neil P. Strickland,** Morava K-theories and localisation, 1999

665 **George Lawrence Ashline,** The defect relation of meromorphic maps on parabolic manifolds, 1999

For a complete list of titles in this series, visit the
AMS Bookstore at **www.ams.org/bookstore/**.